Emitter Detection and Geolocation for Electronic Warfare

For a complete listing of titles in the
Artech House Electronic Warfare Library,
turn to the back of this book.

Emitter Detection and Geolocation for Electronic Warfare

Nicholas A. O'Donoughue

ARTECH HOUSE

BOSTON | LONDON
artechhouse.com

Library of Congress Cataloging-in-Publication Data
A catalog record for this book is available from the U.S. Library of Congress.

British Library Cataloguing in Publication Data
A catalogue record for this book is available from the British Library.

Cover design by John Gomes

ISBN 13: 978-1-63081-564-6

© 2020 ARTECH HOUSE
685 Canton Street
Norwood, MA 02062

All rights reserved. Printed and bound in the United States of America. No part of this book may be reproduced or utilized in any form or by any means, electronic or mechanical, including photocopying, recording, or by any information storage and retrieval system, without permission in writing from the publisher.

All terms mentioned in this book that are known to be trademarks or service marks have been appropriately capitalized. Artech House cannot attest to the accuracy of this information. Use of a term in this book should not be regarded as affecting the validity of any trademark or service mark.

10 9 8 7 6 5 4 3 2 1

To my patient wife, Lauren, who sacrificed many evenings and weekends to give me time to write this book, and to my children, Cecilia and Phineas, who provided countless diversions.

Contents

Preface   xv

Chapter 1   Introduction   1
    1.1   Detection of Threat Emitters   2
    1.2   Angle of Arrival (AOA) Estimation   4
    1.3   Geolocation of Threat Emitters   5
        1.3.1   Geolocation by Satellite   7
    1.4   Signals of Interest   7
    1.5   Nonmilitary Uses   9
    1.6   Limitations   9

I   Detection of Threat Emitters   11

Chapter 2   Detection Theory   13
    2.1   Background   13
        2.1.1   Sources of Variation   13
        2.1.2   Likelihood Function   14
        2.1.3   Sufficient Statistic   15
    2.2   Binary Hypothesis Testing   17
    2.3   Composite Hypothesis   21
    2.4   Constant False-Alarm Rate Detectors   24
        2.4.1   Side Channel Information   25
    2.5   Problem Set   27

Chapter 3 Detection of CW Signals — 29
- 3.1 Background — 29
- 3.2 Formulation — 30
- 3.3 Solution — 33
  - 3.3.1 Threshold Selection — 34
  - 3.3.2 Detection Algorithm — 35
  - 3.3.3 Detection Performance — 36
- 3.4 Performance Analysis — 36
  - 3.4.1 Brief Review of RF Propagation — 37
  - 3.4.2 Detection of an FM Broadcast Tower — 41
  - 3.4.3 CW Radar Detection — 44
- 3.5 Problem Set — 46

Chapter 4 Detection of Spread Spectrum Signals — 49
- 4.1 Background — 50
- 4.2 Formulation — 53
  - 4.2.1 DSSS Encoding — 53
  - 4.2.2 Spread Spectrum Radar Signals — 54
- 4.3 Solution — 54
  - 4.3.1 Energy Detectors — 54
  - 4.3.2 Cyclostationary Detectors — 55
  - 4.3.3 Cross-Correlation Detectors — 58
- 4.4 Performance Analysis — 62
  - 4.4.1 Detection of a 3G CDMA Cell Signal — 64
  - 4.4.2 Detection of a Wideband Radar Pulse — 66
- 4.5 Limitations — 67
- 4.6 Problem Set — 68

Chapter 5 Scanning Receivers — 71
- 5.1 Digital Receivers — 72
  - 5.1.1 In-Phase (I) and Quadrature (Q) Digitization — 73
- 5.2 IF Receivers — 73
- 5.3 Frequency Resolution — 76
- 5.4 Problem Set — 79

II Angle of Arrival Estimation — 83

Chapter 6 Estimation Theory — 85

6.1	Background		85
6.2	Maximum Likelihood Estimation		86
6.3	Other Estimators		88
	6.3.1	Minimum Variance Unbiased Estimators	88
	6.3.2	Bayes Estimators	89
	6.3.3	Least Square Estimators	91
	6.3.4	Convex Estimators	92
	6.3.5	Tracking Estimators	92
6.4	Performance Measures		94
	6.4.1	Root Mean Squared Error (RMSE)	95
	6.4.2	CRLB	95
	6.4.3	Angle Error Variance and Confidence Intervals	101
6.5	Problem Set		103

Chapter 7 Direction-Finding Systems 107

7.1	Beam Pattern-Based Direction Finding		108
	7.1.1	Implementation	110
	7.1.2	Performance	110
7.2	Watson-Watt Direction Finding		117
	7.2.1	Implementation	118
	7.2.2	Performance	119
7.3	Doppler-Based Direction Finding		120
	7.3.1	Formulation	122
	7.3.2	Implementation	123
	7.3.3	Performance	124
7.4	Phase Interferometry		127
	7.4.1	Implementation	129
	7.4.2	Performance	130
	7.4.3	Resolving Ambiguities with Multiple Baselines	132
7.5	Performance Comparison		132
7.6	Monopulse Direction Finding		137
	7.6.1	Performance	139
7.7	Problem Set		141

Chapter 8 Array-Based AOA 145

8.1	Background		145
	8.1.1	Nonstandard Array Configurations	146
8.2	Formulation		148
	8.2.1	Multiple Plane Waves	149

		8.2.2	Wideband Signals	149
		8.2.3	Array Beamforming	149
		8.2.4	Nonisotropic Element Patterns	152
		8.2.5	Gain and Beamwidth	153
		8.2.6	Array Tapers	154
		8.2.7	Two-Dimensional Arrays	157
	8.3	Solution		159
		8.3.1	Signal Models	159
		8.3.2	Maximum Likelihood Estimation	160
		8.3.3	Beamformer Scanning	163
		8.3.4	Subspace-Based Methods	168
	8.4	Performance Analysis		172
		8.4.1	Gaussian Signal Model	172
		8.4.2	Deterministic Signal Model	173
	8.5	Problem Set		178

III Geolocation of Threat Emitters 183

Chapter 9 Geolocation of Emitters 185

- 9.1 Background — 185
- 9.2 Performance Metrics — 186
  - 9.2.1 Error Ellipse — 186
  - 9.2.2 CEP — 190
  - 9.2.3 MATLAB® Code — 193
- 9.3 CRLB — 193
- 9.4 Trackers — 195
- 9.5 Geolocation Algorithms — 195
  - 9.5.1 Ongoing Research — 195
- 9.6 Problem Set — 196

Chapter 10 Triangulation of AOA Measurements 199

- 10.1 Background — 199
- 10.2 Formulation — 200
  - 10.2.1 3-D Geometry — 202
- 10.3 Solution — 203
  - 10.3.1 Geometric Solution for Two Measurements — 203
  - 10.3.2 Geometric Solutions for Three or More Measurements — 203
  - 10.3.3 Maximum Likelihood Estimate — 203

10.3.4 Iterative Least Squares	205
10.3.5 Gradient Descent	207
10.4 Other Solutions	211
10.5 Performance Analysis	211
10.6 Problem Set	214

Chapter 11 TDOA — 217

11.1 Background	217
11.2 Formulation	218
11.2.1 Isochrones	219
11.2.2 Number of Sensors	219
11.3 Solution	220
11.3.1 Maximum Likelihood Estimate	222
11.3.2 Iterative Least Squares Solution	222
11.3.3 Gradient Descent Algorithms	224
11.3.4 Chan-Ho Approach	225
11.3.5 Spherical Methods	227
11.4 TDOA Estimation	228
11.4.1 Time of Arrival Detection	229
11.4.2 Cross-Correlation Processing	230
11.4.3 Clock Synchronization	232
11.5 Geolocation Performance	233
11.6 Limitations	238
11.7 Problem Set	240

Chapter 12 FDOA — 245

12.1 Background	246
12.2 Formulation	247
12.3 Solution	248
12.3.1 Maximum Likelihood Estimate	250
12.3.2 Iterative Least Squares Solution	251
12.3.3 Gradient Descent Algorithms	253
12.3.4 Other Approaches	253
12.4 FDOA Estimation	255
12.4.1 Frequency of Arrival Estimation	255
12.4.2 FDOA Estimation	256
12.4.3 Limitations of Frequency Estimation	257
12.5 Geolocation Performance	257
12.6 Limitations	261

xii　　Contents

　　　　12.7　Problem Set　　263

Chapter 13　Hybrid TDOA/FDOA　　**267**
　　　　13.1　Background　　267
　　　　13.2　Formulation　　268
　　　　13.3　Solution　　270
　　　　　　　13.3.1　Numerically Tractable Solutions　　271
　　　　　　　13.3.2　Other Solutions　　275
　　　　13.4　Joint Parameter Estimation　　277
　　　　　　　13.4.1　AOA Estimation　　277
　　　　　　　13.4.2　Joint Estimation of Time/Frequency Difference　　277
　　　　　　　13.4.3　Full Covariance Matrix　　279
　　　　13.5　Performance Analysis　　280
　　　　13.6　Limitations　　283
　　　　13.7　Problem Set　　283

Appendix A　Probability and Statistics　　**287**
　　　　A.1　Common Distributions　　287
　　　　　　　A.1.1　Gaussian Random Variable　　288
　　　　　　　A.1.2　Complex Gaussian Random Variable　　288
　　　　　　　A.1.3　Chi-Squared Random Variable　　290
　　　　　　　A.1.4　Noncentral Chi-Squared Random Variable　　290
　　　　　　　A.1.5　Rayleigh Random Variable　　292
　　　　　　　A.1.6　Rician Random Variable　　292
　　　　A.2　Student's T Distribution　　293
　　　　A.3　Random Vectors　　294
　　　　　　　A.3.1　Gaussian Random Vectors　　295
　　　　　　　A.3.2　Complex Gaussian Random Vectors　　296

Appendix B　RF Propagation　　**299**
　　　　B.1　Free-Space Propagation　　300
　　　　B.2　Two-Ray Propagation　　300
　　　　B.3　Fresnel Zone　　301
　　　　B.4　Knife-Edge Diffraction　　304
　　　　B.5　Other Models　　305
　　　　B.6　Urban Signal Propagation　　305

Appendix C　Atmospheric Absorption　　**309**
　　　　C.1　Losses Due to Absorption by Gases　　310

	C.2	Losses Due to Absorption by Rain	311
	C.3	Losses Due to Absorption by Clouds and Fog	313
	C.4	Standard Atmosphere	313
	C.5	Wrapper Function	314
	C.6	Zenith Attenuation	314
	C.7	MATLAB® Toolboxes and Model Fidelity	315

Appendix D System Noise   **317**

	D.1	Additive White Gaussian Noise	317
	D.2	Colored Noise	319
	D.3	Sky Noise	319
		D.3.1 Cosmic Noise	320
		D.3.2 Atmospheric Noise	323
		D.3.3 Ground Noise	323
	D.4	Urban (Man-Made) Noise	324

About the Author   **327**

Index   **329**

Preface

The field of electronic warfare is very broad, and research is still active. A quick glance at the catalog of already published works reveals no shortage of introductory texts, historical accounts, and expertly researched manuals. This book aims to strike a balance between clearly written introductory texts, such as David Adamy's *EW 101* series of books, and comprehensive reference manuals, such as Richard Poisel's *Electronic Warfare: Receivers and Receiving Systems*. Throughout this book, references are provided for additional reading.

This book provides a consistent theoretical formulation for detection and geolocation of threat emitters. It contains a formal background, rooted in statistical signal processing and array processing theory. The book is broken into Parts I-III, titled, respectively, "Detection of Threat Emitters," "Angle of Arrival Estimation," and "Geolocation of Threat Emitters."

MATLAB® code for the examples given, figures generated, and algorithms presented is provided on the Artech House website. Problem sets are provided for each technical chapter. Once downloaded, the code can be run from the root folder of the package or by adding that root folder to the MATLAB path.

The consistent formulation is useful for college courses, at the graduate or advanced undergraduate level, supported by the problem sets. The book's MATLAB code assists practitioners, system designers, and researchers in understanding and implementing the algorithms in the text, as well as developing new approaches to solving these problems.

The target audience has some background in probability and statistics and in signal processing concepts, such as the *Nyquist sampling criteria* and *Fourier transform*.

Chapter 1

Introduction

The electromagnetic spectrum is used heavily on the modern battlefield. One of the principal functions of *electronic warfare* (EW) is to detect these emissions and determine their point of origin. Detection serves several crucial purposes, such as alerting a pilot that an adversary radar has established a track (and is therefore ready to fire a missile), or detecting the presence of enemy jamming. Determining the location of an emitter allows one to maneuver towards or away from the threat, cue GPS-enabled munitions or off-board jamming support, or even just determine the disposition of enemy forces, for example.

Detection and geolocation of threat emitters has been a critical piece of military combat for more than a century. Early *radiofrequency* (RF) direction-finding systems were patented in 1906 and rapidly improved in the decades leading up to World War II [1]. As part of the preparation for the Normandy invasion, direction-finding stations on the British coast were used to detect and locate German coastal defense radars, which were then targeted with aerial bombardments. The ability to pinpoint German radar positions allowed the Allied forces to first destroy as many as possible, and then later to construct an elaborate jamming campaign that masked the approach of the Allied invasion fleet and deceived the surviving radar operators into thinking fleets were approaching different beaches. This gave the Allied invasion enough cover to make landfall before the Germans could concentrate their forces to repel them [2]. The utility of electronic warfare has only grown in the decades since, as modern military forces have become increasingly dependent on RF emissions to observe and communicate.

Today, *radar warning receivers* (RWRs) routinely provide pilots with detection and geolocation of threats, and increasingly capable integrated EW systems are pushing the bounds in terms of sensitivity and accuracy of detections and communication of that information to other platforms for geolocation processing and engagement. The U.S military has experienced a resurgence in funding and development of EW systems, particularly in the area of battle management and situational awareness of the electromagnetic spectrum.[1] A key requirement for that situational awareness is information about the adversary's disposition of forces, provided by detection and geolocation assets.

The text is broken into Parts I–III: Detection of Threat Emitters, Angle of Arrival Estimation, and Geolocation of Threat Emitters. Each part begins with an introductory chapter covering the relevant signal processing theory and then provides a series of relatively short chapters covering specific methods to achieve the desired end product. Sections 1.1-1.3 briefly discuss Parts I–III.

1.1 DETECTION OF THREAT EMITTERS

Many different detection architectures exist, each with pros and cons. Some of the more popular receiver architectures include the scanning superheterodyne receiver, which involves a narrowband receiver that is scanned over a very wide bandwidth; compressive receivers, which assign frequency channels to successive temporal bins; and channelized receivers, which involve a very wide bandwidth digitization and subsequent separation of the signals into different digital channels. Each of the receivers has advantages and limitations, and the choice of optimal architecture will depend very much on the characteristics of the target signals, such as bandwidth and spread spectrum characteristics.

In reality, all collected signals are stochastic in nature. This is chiefly due to the presence of noise, the majority of which comes from galactic sources or thermal background noise, as well as inefficiencies in RF hardware. The typical approach to dealing with these sources is to model them, in aggregate, as Gaussian noise. When expected signal levels are calculated, they are not deterministic; there is a probability distribution attached, and the reported signal level is often just the expected value. Often the relevant metric is the *signal-to-noise ratio* (SNR), which is a measure of the signal power divided by the noise power.

When we define a detection threshold in reference to some maximum tolerated probability of false alarm (the chance that noise alone can cause the signal to

---

1 This was largely caused by the 2013 Defense Science Board Summer Study on Electronic Warfare, which recommended a significant push to improve both the technology and approach to fielding EW systems for the U.S. military. A summary is available to the public [3].

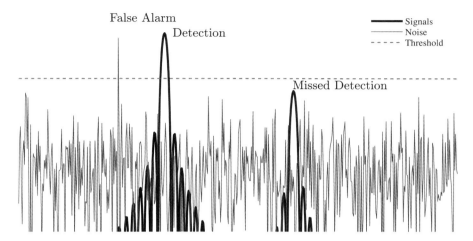

Figure 1.1 A proper detection threshold must be set to minimize the likelihood of a missed detection, without allowing for excessive false alarms.

exceed the threshold) and the desired probability of detection (the probability that the desired signal will exceed the threshold). Figure 1.1 depicts a typical detection scenario with two signals to detect in the presence of noise. The detection threshold is shown as a dotted line; one of the signals is detected, while the other is missed, and a false alarm is registered at one point.

Readers interested in a basic introduction to detection of signals for EW should consult one of Dave Adamy's many well-written texts, such as *EW 103: Tactical Battlefield Communications Electronic Warfare* [4]. For a more thorough treatment of the various receiver architectures, interested readers are directed to the voluminous reference *Electronic Warfare Receivers and Receiving Systems* [5]. Readers interested in the mathematics of detection theory, on which the formulations in this text are based, should consult Louis Scharf's *Statistical Signal Processing* [6].

Detection approaches have a number of key performance parameters; most notable to this text are the false alarm rate (false alarms/hour) of the detector, minimum detectable signal (for a given probably of detection), and probability of detection (for some given observation time). For detection of emitted signals, it is important to analyze the receiver's architecture (is it staring at a wide bandwidth, sweeping a narrow receiver bandwidth, or using a compressive mixer) as well as the characteristics of the signal being observed (bandwidth, frequency hopping, and pulse duration, for example), to determine detection performance, which can be stated in terms such as the probability of a single pulse or transmissions being detected, or the expected delay until a detection occurs.

Detection theory, and performance predictions, are discussed more fully in Chapter 2. Chapter 3 discusses the detection of continuous wave (CW) signals, and Chapter 4 discusses spread spectrum signals. The implications of a scanning receiver, which must sample a much larger bandwidth than the signal is occupying, are discussed in Chapter 5.

1.2 ANGLE OF ARRIVAL (AOA) ESTIMATION

AOA estimation provides an incremental improvement in situational awareness (SA), namely the information that an adversary can be found in a certain direction. It can also be used to cue some responses, such as jamming the emitter, or to launch and guide a missile. Without range information, it is difficult to assess how effective an electronic attack will be, and missile guidance is degraded when fired on an angle-only track, resulting in a reduced likelihood of successful engagement. Nevertheless, AOA is often the best that can be achieved from a single platform and provides some benefit.

Similar to detection, direction finding of RF emitters can occur in a number of different ways, each with its own set of advantages and limitations. We refer to the solution as AOA estimation, but it is also commonly referred to as a *line of bearing* (LOB). The estimate can be accomplished using the directional characteristics of individual antennas, complex arrangements of closely spaced antennas, or even a phased array, in increasing order of both complexity and accuracy [1, 7].

An illustration of AOA estimation is shown in Figure 1.2; the solid line shows the estimated AOA, and the shaded region indicates the uncertainty of that estimate. The cause of this uncertainty includes the stochastic noise sources that also impact detection as well as the interactions between multiple emitters, which can cause distortions and errors in the AOA algorithms.

Direction-finding performance is quantified most significantly by the *standard deviation* of the AOA estimate, and the *angular resolution*, which is the closest angular separation between two emitters that can be tolerated before the system is unable to resolve them. Estimation theory, including these performance metrics and others, is discussed in Chapter 6, and basic methods of direction finding are discussed in Chapter 7. Array-based techniques are outlined in Chapter 8.

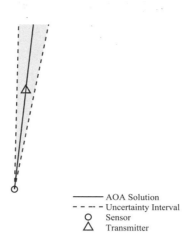

Figure 1.2 AOA Estimation of an emitter. The shaded region shows the uncertainty interval, in this case ±5 deg.

1.3 GEOLOCATION OF THREAT EMITTERS

Geolocation is the estimation of an emitter's position in absolute terms, either global coordinates such as latitude and longitude, or local coordinates, such as range and bearing from a fixed reference. One approach to geolocation is to generate lines of bearing from multiple sensors and combine them in a process called *triangulation* to form a geolocation estimate as shown in Figure 1.3. Triangulation can also be conducted from a single moving platform (if the emitter is sufficiently slow compared to the sensing platform), typically either an aircraft or a satellite in low Earth orbit [4, 7, 8].

The use of multiple stations also brings the ability to do precision geolocation via techniques such as the time difference of arrival (TDOA) or frequency difference of arrival (FDOA) at multiple sites, or from multiple platforms (aircraft or spacecraft) [9]. When considered together, TDOA and FDOA can enable formation of a geolocation estimate with a single pair of sensors, under certain limitations. TDOA is discussed in Chapter 11; FDOA in Chapter 12, and hybrid TDOA/FDOA in Chapter 13.

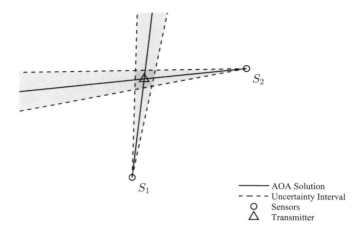

Figure 1.3 Emitter geolocation by triangulation from two sensor AOA estimates. Shaded regions show the uncertainty intervals on the individual measurements, and the intersection of the solutions shows the uncertainty in the estimated emitter position.

The most prominent performance metric for geolocation is the *circular error probable* (CEP), defined as the radius of a circle around the estimated position, within which the target has a specified chance of residing (typically either 50% or 95%). Other measures of performance include statistical bounds on performance, such as the *Cramer-Rao lower bound* (CRLB), which provides a lower bound on the mean-squared error of an *unbiased* estimator,[2] or the Ziv-Zakai lower bound (ZZB), which provides a tighter (more accurate) performance bound in low to moderate SNR regimes. Geolocation performance is discussed in detail in Chapter 9.

For background reading on geolocation, interested readers are again referred to *EW 103* [4]. A more complete treatment of geolocation is found in Poisel's *Electronic Warfare Target Location Methods* [9], which introduces many of the geolocation methods discussed in Parts II and III of this book.

---

2 An unbiased estimator is one in which the estimates are centered on the true value (i.e., there is no *bias* in the estimates).

1.3.1 Geolocation by Satellite

Satellites are a good choice for collection of *electronic intelligence* (ELINT), including the estimation of emitter location, partially because they can overfly any country in the world without provoking a military response, providing access to signals that would otherwise by impossible to detect. The United States launched the first ELINT satellite in the 1960s and has since been joined by several other countries [8]. With the advent of smaller satellites, often called *cubesats* or *pico satellites,* it is now possible for even individuals and academic institutions to launch experimental payloads [10, 11].

The geolocation performance of satellites is directly impacted by their long range to the emitters, but their high speed and predictable orbits enable the use of geolocation techniques that do not make sense for land-based or aerial detectors, including several single-ship techniques. For a detailed description of satellite-based geolocation, the interested reader is referred to *Space Electronic Reconnaissance: Localization Theories and Methods* [8].

1.4 SIGNALS OF INTEREST

There is a variety of potential signals that one might wish to detect and geolocate. In the context of EW, the principal signals of interest are radar and communications signals.

Radar signals vary greatly by application; carrier frequencies range from HF (3–30 MHz) for over-the-horizon search radars, to millimeter-wave (30–300 GHz) for missile seekers and wideband satellite imaging. Signal bandwidths are at least 1 MHz and can exceed 1 GHz for high-resolution imaging applications. *Pulse repetition intervals* (PRIs) can be as slow as a few milliseconds for very long-range systems or as fast as a few microseconds for short-range tracking radars. Radar signals can be very unpredictable, and a single radar can rapidly change modes as it performs different tasks. Power levels vary greatly by mission and platform, but typically exceed 30 dBW (1 kW) for manned airborne platforms and 60 dBW (1 MW) for shipborne and land-based systems. Most important, however, is the fact that most radars are highly directional, meaning that a receiver that is not being illuminated by the radar will have to be capable of detecting sidelobe emissions, which can be as much as 40 dB weaker than the mainlobe.

While legacy radar systems were often characterized by fixed waveforms and behaviors, the advent of solid-state amplifiers and active electronically scanned arrays (AESAs), has greatly increased the flexibility and unpredictability of radar systems. The overriding feature of a radar signal is that it may be wholly unpredictable from one *coherent processing interval* (CPI) to the next, sometimes even from one pulse to the next.

Communications signals are largely more structured. In order for two systems to exchange information, they must agree on how that information is coded, modulated, and transmitted through the electromagnetic spectrum. Multiple transmitters are deconflicted by separating them in time, frequency, and space, or through a time-frequency code. Communication signals are one-way, while radar signals must reflect off a target and repropagate to the receiver. For that reason, communication signals are typically much weaker, less than 30 dBW (1 kW) in almost all circumstances, with the exception of some satellite transmission systems. The last significant difference is that communications signals are frequently, although not always, transmitted in all directions.

The principal challenge for detection of military waveforms is the use of frequency hopping. While the channels that are available are often either publicly known or easy to decipher (since they are unlikely to change over time), the channel at any given moment is selected at random (according to a secret key known only to the network participants), and these frequencies are changed very rapidly. In the case of Link-16,[3] the hop rate is more than 16,000 hops per second, meaning that the signal dwells at any one frequency for less than 100 μs. This presents a challenge for EW receivers, which must be prepared to instantaneously sample the entire bandwidth of available signals in order to acquire the signal after it has hopped to a new frequency, and the less time the signal spends at each frequency, the harder it is for an EW system to gather enough of the signal to detect and process. A related technique often used in military systems is *direct-sequence spread spectrum* (DSSS), which takes a relatively narrow bandwidth signal and modulates it with a repeating code based on a secret key. This has the effect of spreading the energy over a wider swath of RF spectrum, potentially reducing the signal density below the noise floor. This significantly complicates the effort necessary to detect such a signal.

In addition to communication signals, there is also a number of transponders that may be of interest to detect and locate. These include the *automated information*

---

3   Link-16 is a U.S. military datalink that is widely used among U.S. and allied aircraft, and some ground users. It is an L-band pulsed waveform, with frequency hopping and encryption citengl16guide.

system (AIS), which is mandated for cargo vessels and voluntarily used by many commercial fishing vessels, and operates on a fixed frequency (1.612 GHz) with a small set of possible messages. All vessels are issued a unique ID and broadcast their coordinates (typically GPS-derived) to any receivers in the area, as well as satellite receivers in orbit. Aircraft have a similar transponder system, called *automated dependent surveillance-broadcast* (ADS-B). Detection and location of these signals can be an effective way to detect systems that are broadcasting an incorrect position (either due to equipment malfunction or to intentional deception). Most military aircraft have similar systems, called *integrated friend or foe* (IFF) transponders. While they are typically encrypted and their details classified, they do represent an attractive target for detection and geolocation.

1.5 NONMILITARY USES

The techniques discussed in this book are not restricted to military EW systems. There are many non military uses of these techniques, some of which have unique characteristics that impact the performance of various techniques. For instance, these techniques are useful for source localization, such as to detect gas leaks or natural phenomenon based on measurements from a sensor network, or tracking the movement of wildlife. Similar techniques are used to perform indoor position location of users, such as robots navigating a factory floor, or customers navigating a shopping mall. In indoor and urban scenarios, tremendous care must be taken to account for the impact of multipath echoes. Cell phone providers can use similar techniques to estimate the position of all of their users; while a coarse position is provided by the tower one accesses, often signals are received by multiple towers, enabling them to estimate a user's position with great accuracy, even in the absence of GPS, and without cooperation from the user's cell phone [12].

1.6 LIMITATIONS

One of the principal limitations of emitter detection and geolocation is its inherent reliance on emissions. Systems that want to avoid being subject to electronic reconnaissance exercise *emissions control* (EMCON). The cost of this, of course, is that mission effectiveness is often hindered by EMCON, so many systems seek to balance the risk of detection against the benefit of operating in the RF spectrum.

Bistatic radar is a challenge for emitter geolocation, in that only the transmitter is observable. While this allows for collection of the signal for intelligence purposes, and location of the transmitter for kinetic engagement, it leaves the receiver station undetected, which defeats any effort to cue jamming support against the detected radar. A special class of bistatic radar is passive radar, which does not use any transmitter whatsoever. These systems rely on emissions by known (and typically stationary) emitters in the vicinity, which defeats any attempt to locate them by electromagnetic means. Typically these systems rely on high-power broadcast stations, such as radio or television transmitters [13].

References

[1] T. E. Tuncer and B. Friedlander, *Classical and modern direction-of-arrival estimation.* Boston, MA: Academic Press, 2009.

[2] H. Griffths, "The d-day deception operations taxable and glimmer," *IEEE Aerospace and Electronic Systems Magazine*, vol. 30, no. 3, pp. 12–20, 2015.

[3] D. S. Board, "21st century military operations in a complex electromagnetic environment," Office of the Secretary of Defense, Tech. Rep. AD1001629, July 2015.

[4] D. Adamy, *EW 103: Tactical battlefield communications electronic warfare.* Norwood, MA: Artech House, 2008.

[5] R. A. Poisel, *Electronic warfare receivers and receiving systems.* Norwood, MA: Artech House, 2015.

[6] L. L. Scharf, *Statistical signal processing.* Reading, MA: Addison-Wesley, 1991.

[7] A. Graham, *Communications, Radar and Electronic Warfare.* Hoboken, NJ: John Wiley & Sons, 2011.

[8] F. Guo, Y. Fan, Y. Zhou, C. Xhou, and Q. Li, *Space electronic reconnaissance: localization theories and methods.* Hoboken, NJ: John Wiley & Sons, 2014.

[9] R. Poisel, *Electronic warfare target location methods.* Norwood, MA: Artech House, 2012.

[10] H. Heidt, J. Puig-Suari, A. Moore, S. Nakasuka, and R. Twiggs, "Cubesat: A new generation of picosatellite for education and industry low-cost space experimentation," in *Proceedings 2000 Small Satellite Conference*, 2000.

[11] K. Woellert, P. Ehrenfreund, A. J. Ricco, and H. Hertzfeld, "Cubesats: Cost-effective science and technology platforms for emerging and developing nations," *Advances in Space Research*, vol. 47, no. 4, pp. 663–684, 2011.

[12] R. Zekavat and R. M. Buehrer, *Handbook of position location: Theory, practice and advances.* Hoboken, NJ: John Wiley & Sons, 2011, vol. 27.

[13] M. A. Richards, J. Scheer, W. A. Holm, and W. L. Melvin, *Principles of Modern Radar.* Edison, NJ: SciTech Publishing, 2010.

Part I

Detection of Threat Emitters

Chapter 2

Detection Theory

This chapter discusses detection theory, one of the principal arms of statistical signal processing, sometimes referred to as *hypothesis testing* or *decision making*. The basic premise of detection theory is that the detector has access to a noisy measurement, and the goal is to determine whether or not a target signal is present in those measurements. This chapter discusses the basics of detection theory, as they apply to emitter detection. Chapters 3 and 4 discuss application of this theory to signals of interest, including continuous and spread spectrum signals, respectively. Chapter 5 discusses the various receiver architectures that are employed to observe a broad swath of frequencies for any signals that may be present.

This chapter is necessarily an introduction to the field. Interested readers are referred to seminal texts by Louis Scharf [1], Harry Van Trees [2], or Steven Kay [3].

2.1 BACKGROUND

This chapter assumes some familiarity with random variables, common statistical distributions, and stochastic processes. Some of this background material is presented in Appendix A.

2.1.1 Sources of Variation

There are myriad sources of variation in received signals, many of which are referred to as noise or interference, for the sole reason that they are not the signal of interest. Sources of noise and interference include RF signals emitted by celestial bodies, referred to as *galactic background noise,* as well as emissions that arise from RF circuitry and components, referred to as *thermal noise,* and man-made emissions, such as RF broadcasts and intentional jamming signals.

Mathematically, this is expressed with the equation

$$N = kTB \qquad (2.1)$$

where k is Boltzmann's constant ($1.38 \times 10^{-23} W \cdot s/K$), T is the device temperature, and B is the receiver bandwidth (in Hertz). This does not include man-made interference, or system inefficiencies. The standard temperature that is typically referenced is $T = 290K$ (room temperature), which results in a noise power spectral density of -144 dBW/MHz (i.e., for every 1 MHz of receiver bandwidth, -144 dBW of noise enters the system). The most significant contributing factor to noise is often the nonideal behavior of RF components in the receiver, which amplify incoming noise. The performance of these components is measured and expressed as a *noise factor*. The total effective noise in the system is thus given as[1]

$$N = kTBF_N \qquad (2.2)$$

where F_N is the *noise factor*, sometimes called the *noise figure*. The calculation of noise power levels is discussed in more detail in Appendix D.

2.1.2 Likelihood Function

A set of measurements z, taken over time or from an array of sensors, is affected by unknown parameters, which we denote with the term θ. The parameterized likelihood function is written $f_\theta(\mathbf{z})$, which represents the probability that a measurement z is observed given the parameter vector θ.

Figure 2.1 shows an example of a likelihood function $f_\theta(\mathbf{z})$ with three possible values of the unknown parameter θ: θ_1, θ_2, and θ_3. In Figure 2.1, θ serves to shift the mean value of the measurement z, although in other cases it could modify the variance in addition to or instead of the mean, or it could introduce completely different distribution functions (i.e., Rician for one value, Gaussian for another...).

In most emitter detection situations, the unknown parameter θ represents the impact of attenuation and delay as a signal propagates some unknown distance through an imperfectly measured atmosphere.

---

1 Some references use a slightly different definition of the noise factor, $N = kTB(F_N - 1)$.

Detection Theory

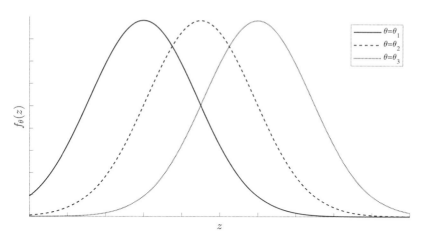

Figure 2.1 An example scenario where the parameter θ has three possible values, which affect the mean value of the measurement z.

2.1.3 Sufficient Statistic

While the measurement vector **z** provides some information about the unknown parameter **θ**, it is often difficult to construct a decision rule that uses the full measurement vector. A *sufficient statistic* provides a lower dimensional (often scalar) signal $T(\mathbf{z})$ that is equally useful for estimating **θ**. To find, or prove the existence of, a sufficient statistic, the likelihood function must be separable into two functions, one that depends on the sufficient statistic and the unknown parameter, and one that contains any remaining dependence on **z**, but is independent of θ. In other words, if a representation

$$f_\theta(\mathbf{z}) = a(\mathbf{z}) b_\theta\left(T(\mathbf{z})\right) \qquad (2.3)$$

can be found, then $T(\mathbf{z})$ represents a sufficient statistic for **z** when estimating the parameter θ, and it can be used without any loss of performance.

Example 2.1 Sample Mean

As an example, consider a complex Gaussian random vector **z** whose N elements are independent and identically distributed (IID) with known variance σ_n^2 and unknown mean θ.[2] The likelihood function is given by

$$f_\theta(\mathbf{z}) = \frac{1}{\pi^N \sigma_n^{2N}} e^{-\frac{1}{\sigma_n^2} \sum_{i=1}^N |z_i - \theta|^2} \qquad (2.4)$$

We expand the square in the exponent, and break apart the summation into three different exponentials, to yield an expanded likelihood function[3]

$$f_\theta(\mathbf{z}) = \frac{1}{\pi^N \sigma_n^{2N}} e^{-\frac{1}{\sigma_n^2} \sum_{i=1}^N |z_i|^2} e^{\frac{2}{\sigma_n^2} \sum_{i=1}^N \Re\{\theta^* z_i\}} e^{-\frac{N}{\sigma_n^2} |\theta|^2} \qquad (2.5)$$

This allows us to collect all of the terms that are independent of θ into the function $c(\mathbf{z})$ and rewrite the likelihood function as

$$f_\theta(\mathbf{z}) = c(\mathbf{z}) e^{\frac{2}{\sigma_n^2} \sum_{i=1}^N \Re\{\theta^* z_i\}} e^{-\frac{N}{\sigma_n^2} |\theta|^2} \qquad (2.6)$$

$$= c(\mathbf{z}) \underbrace{e^{\frac{2}{\sigma_n^2} \Re\{\theta^* T(\mathbf{z})\}} e^{-\frac{N}{\sigma_n^2} |\theta|^2}}_{b_\theta(T(\mathbf{z}))} \qquad (2.7)$$

where $T(\mathbf{z}) = \sum_{i=1}^N z_i$. This satisfies the condition above for a sufficient statistic. For convenience, we apply a scalar multiplication term, to arrive at the sample mean

$$T(\mathbf{z}) = \frac{1}{N} \sum_{i=1}^N z_i \qquad (2.8)$$

Similarly, if the desired signal is Gaussian with a known mean (μ) and unknown variance (θ), it can be shown that the sample variance

$$T(\mathbf{z}) = \frac{1}{N-1} \sum_{i=1}^N |z_i - \mu|^2 \qquad (2.9)$$

is a sufficient statistic for the unknown variance (θ).

---

[2] The complex Gaussian distribution represents a complex variable z whose real and imaginary components are jointly Gaussian. The distribution is discussed in more detail in Appendix A.
[3] $\Re\{\cdot\}$ is the real operator and returns the real component of a complex number.

2.2 BINARY HYPOTHESIS TESTING

In mathematical terms, detection is often referred to as *hypothesis testing*, in that a detection system is attempting to test two or more hypotheses regarding whether a signal is present or absent in a given set of measurements, and what its parameters are (amplitude, phase, etc.).

In its simplest form, detection is expressed as a binary hypothesis. We first define the two hypotheses that will be tested, commonly referred to as the *null hypothesis* (signal is absent) and *alternative hypothesis* (signal is present). We denote these hypotheses with the terms $\mathcal{H}_0$ and $\mathcal{H}_1$. We also need a measurement model for the received signal vector z. We define the signal model

$$\mathbf{z} = \theta \mathbf{s} + \mathbf{n} \quad (2.10)$$

where s is the underlying signal of interest and n is the noise vector. Noise is frequently assumed to be white (spectrally flat) and Gaussian with variance σ_n^2, thus the covariance matrix for a noise vector n is given $\sigma_n^2 \mathbf{I}$. This isn't always realistic, but is sufficiently accurate for most purposes, and is a widely accepted assumption. The parameter θ can then be expressed in the two hypotheses with

$$\theta = \begin{cases} \theta_0 & \mathcal{H}_0 \\ \theta_1 & \mathcal{H}_1 \end{cases} \quad (2.11)$$

This notation indicates that under the null hypothesis (signal is absent), $\theta = \theta_0$ (typically 0), and under the alternative hypothesis (signal is present), $\theta = \theta_1$.

The next question to address, then, is how to determine an appropriate test. One straightforward approach is to declare the *most likely* scenario to be true. This is accomplished via a calculation known as the *likelihood ratio test*, defined simply as the ratio of the likelihood function $f_\theta(\mathbf{z})$ under the two hypotheses. Equivalently, we often use the log-likelihood ratio:

$$\ell(\mathbf{z}) = \ln\left(\frac{f_{\theta_1}(\mathbf{z})}{f_{\theta_0}(\mathbf{z})}\right) \quad (2.12)$$

This is then compared to some threshold η. If $\ell(\mathbf{z}) \geq \eta$, then we declare that $\mathcal{H}_1$ is true; otherwise we declare that $\mathcal{H}_0$ is true. In other words, we compute how likely it is that the data vector z would have been received under the assumption that $\mathcal{H}_1$ is true, and compare that to how likely it is that z would have been received if $\mathcal{H}_0$ were true. If this ratio is greater than η, then $\mathcal{H}_1$ is sufficiently more likely to be true than $\mathcal{H}_0$ is, and a detection is declared.

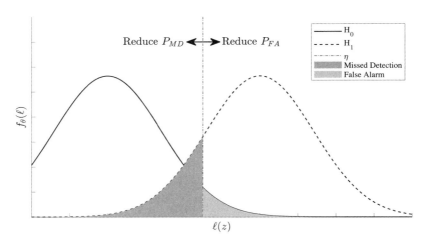

Figure 2.2 Illustration of detection errors; false alarms (light grey) and missed detections (dark grey). Moving the threshold can improve one type of error, at the expense of increasing the other. Proper detector design is a balance between the two.

To choose an appropriate threshold η, we must first analyze the possible errors. Consider Figure 2.2. The solid curve denotes the likelihood ratio under $\mathcal{H}_0$, while the dashed curve indicates the likelihood ratio under $\mathcal{H}_1$. If the measured likelihood ratio is to the right of the threshold, then the system declares that $\mathcal{H}_1$ is true. However, there is a chance that an error was made, and $\mathcal{H}_0$ is actually true. This is referred to as a *false alarm*, and the probability of this occurring can be computed by integrating the area of the likelihood ratio under $\mathcal{H}_0$ that is to the right of the threshold

$$P_{\text{FA}} = Pr_{\theta_0}\left[\ell(\mathbf{z}) \geq \eta\right] = \int_{\eta}^{\infty} f_{\theta_0}(t)dt \qquad (2.13)$$

This is shown as the light grey region in Figure 2.2. Similarly, a declaration of $\mathcal{H}_0$ when $\mathcal{H}_1$ is actually true is referred to as a *missed detection* and is represented by the area of the likelihood ratio under $\mathcal{H}_1$ that is to the left of the threshold.

$$P_{\text{MD}} = Pr_{\theta_1}\left[\ell(\mathbf{z}) < \eta\right] = \int_{-\infty}^{\eta} f_{\theta_1}(t)dt \qquad (2.14)$$

This is shown as the dark grey region in Figure 2.2. By selecting a larger threshold, one can reduce the likelihood of false alarm, but the cost is an increased likelihood of a missed detection. Similarly, by selecting a smaller threshold, one can reduce the probability of a missed detection at the cost of additional false alarms.

The most common approach is to fix a desired P_{FA} and use (2.13) to compute the corresponding threshold η.[4]

Example 2.2 Gaussian Random Vector

Consider the test

$$\begin{aligned} \mathcal{H}_0 &: \quad \mathbf{z} \sim \mathcal{CN}\left(\boldsymbol{\theta}_0, \sigma_n^2 \mathbf{I}\right) \\ \mathcal{H}_1 &: \quad \mathbf{z} \sim \mathcal{CN}\left(\boldsymbol{\theta}_1, \sigma_n^2 \mathbf{I}\right) \end{aligned} \qquad (2.15)$$

Derive the optimal detector and analyze its performance.[5]

For simplicity, we introduce a vectorized form of the complex Gaussian PDF.[6]

$$f_{\boldsymbol{\theta}}(\mathbf{z}) = \pi^{-N} \sigma_n^{-2N} \exp\left\{ \frac{-1}{\sigma_n^2} (\mathbf{z} - \boldsymbol{\theta})^H (\mathbf{z} - \boldsymbol{\theta}) \right\} \qquad (2.16)$$

We compute the log likelihood ratio by plugging (2.16) into (2.12) for both $\mathcal{H}_1$ and $\mathcal{H}_0$

$$\ell(\mathbf{z}) = c + \Re\left\{ \frac{2}{\sigma_n^2} (\boldsymbol{\theta}_1 - \boldsymbol{\theta}_0)^H \mathbf{z} + \frac{1}{\sigma_n^2} (\boldsymbol{\theta}_1 + \boldsymbol{\theta}_0)^H (\boldsymbol{\theta}_0 - \boldsymbol{\theta}_1) \right\} \qquad (2.17)$$

where the term c collects all of the constants in the log likelihood ratio that are independent of the input $\mathbf{z}$ and can be ignored. For simplicity, we define the terms

$$\mathbf{w} = \frac{1}{\sigma_n^2} (\boldsymbol{\theta}_1 - \boldsymbol{\theta}_0) \qquad (2.18)$$

$$\boldsymbol{\theta}^* = \frac{1}{2} (\boldsymbol{\theta}_1 + \boldsymbol{\theta}_0) \qquad (2.19)$$

$$d^2 = \sigma_n^2 \mathbf{w}^H \mathbf{w} \qquad (2.20)$$

and rewrite the log likelihood ratio

$$\ell(\mathbf{z}) = \Re\left\{ \mathbf{w}^H (\mathbf{z} - \boldsymbol{\theta}^*) \right\} \qquad (2.21)$$

---

[4] This is a result of the *Neyman-Pearson lemma* and is one criterion for optimality of a detector. This lemma is discussed in detail in [1], along with other detector types, such as *Bayesian* and *maximum a posteriori* (MAP) detectors.

[5] The symbology $\sim$ stands for "is distributed as," and the $\mathcal{CN}$ indicates a complex Gaussian random vector, parameterized by the expectation and the covariance matrix. See Appendix A for details.

[6] $(\cdot)^H$ is the *Hermitian transpose*, also referred to as the conjugate transpose, and represents both a matrix transpose and complex conjugate operation on each element.

The test statistic $\ell(\mathbf{z})$ is the real component of a linear combination of complex Gaussian random variables (the elements of $\mathbf{z}$), so it is distributed as a Gaussian random variable. Thus, we can re-express the binary hypothesis problem

$$\begin{aligned} \mathcal{H}_1 &: \ell(\mathbf{z}) \sim \mathcal{N}\left(\tfrac{d^2}{2}, d^2\right) \\ \mathcal{H}_0 &: \ell(\mathbf{z}) \sim \mathcal{N}\left(-\tfrac{d^2}{2}, d^2\right) \end{aligned} \quad (2.22)$$

Recall the P_{FA} from (2.13), and note that it is the inverse of the cumulative distribution function (CDF) under $\mathcal{H}_0$. The CDF of a Gaussian random variable with mean value μ and variance σ_n^2 is

$$F(x) = \frac{1}{2}\left(1 + \operatorname{erf}\left(\frac{x-\mu}{\sigma_n\sqrt{2}}\right)\right) \quad (2.23)$$

where $\operatorname{erf}(x)$ is the *error function*, defined

$$\operatorname{erf}(x) = \frac{2}{\sqrt{\pi}}\int_0^x e^{-t^2}\, dt \quad (2.24)$$

and μ and σ_n^2 are the mean and variance, respectively. Thus,

$$\begin{aligned} P_{\text{FA}} &= 1 - F(\eta) & (2.25) \\ &= \frac{1}{2}\left(1 - \operatorname{erf}\left(\frac{\eta + d^2/2}{d\sqrt{2}}\right)\right) & (2.26) \end{aligned}$$

To solve for η, we merely invert (2.26)

$$\eta = d\sqrt{2}\,\operatorname{erfinv}\left(1 - 2P_{\text{FA}}\right) - \frac{d^2}{2} \quad (2.27)$$

where $x = \operatorname{erfinv}(y)$ is the parameter x that satisfies $y = \operatorname{erf}(x)$.

Once a threshold η is selected, we can repeat the process using the PDF of $\ell(\mathbf{z})$ under $\mathcal{H}_1$ to compute the probability of detection (P_D)

$$P_D = \frac{1}{2}\left(1 - \operatorname{erf}\left(\frac{\eta - d^2/2}{d\sqrt{2}}\right)\right) \quad (2.28)$$

To further illustrate the relationship between P_{FA} and P_D, we can plug the solution for the η in (2.27) into (2.28)

$$P_D = \frac{1}{2}\left(1 - \operatorname{erf}\left(\operatorname{erfinv}\left(1 - 2P_{\text{FA}}\right) - d/\sqrt{2}\right)\right) \quad (2.29)$$

For a given P_{FA} the achievable P_D is given by the distance d^2 of the two distributions. This distance term measures how much overlap there is between the distributions of $\ell(\mathbf{z})$ under the two hypotheses. The less overlap there is, the better the detector can perform for a given P_{FA}.

We plot this relationship for a few representative values of d^2 in Figure 2.3. This plot is often referred to as a *receiver operating characteristics* (ROC) curve. The following is MATLAB® code used to generate this plot.

```
% Set up PFA and SNR vectors
pfa = linspace(1e-9,1,1001);
d2_vec = [1e-6,1,5,10];      % d^2, as defined in (2.20)
[PFA,D2] = ndgrid(pfa,d2_vec); % Use nd-grid to clean up code

% Compute the threshold according to (2.25)
eta = sqrt(2*D2).*erfinv(1-2*PFA)-D2/2;

% Compute the probability of detection
PD = .5*(1-erf((eta-D2/2)./sqrt(2*D2)));
```

Alternatively, to compute the ROC in a more compact fashion, compute (2.29) directly, instead of first computing the threshold η to yield:

```
PD = .5*(1-erf(erfinv(1-2*PFA)-D/sqrt(2)));
```

2.3 COMPOSITE HYPOTHESIS

Section 2.2 assumed that the unknown parameter θ took on one of two known values. In reality, they are often unknown. The most common example is with unknown signal strength (due to a whole host of factors, such as unknown distance, atmospheric conditions, and radiated power). This is known as a *one-sided composite hypothesis test*

$$\begin{aligned} \mathcal{H}_0 &: \quad \theta = \theta_0 \\ \mathcal{H}_1 &: \quad \theta > \theta_0 \end{aligned} \quad (2.30)$$

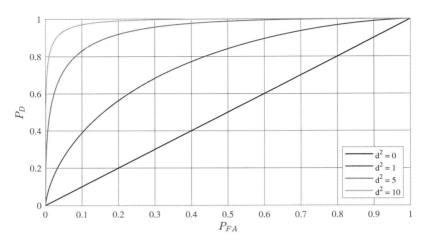

Figure 2.3 ROC for a Gaussian test with varying distances (d^2) between the $\mathcal{H}_0$ and $\mathcal{H}_1$ likelihood functions, as described in (2.29).

In this case, we can't construct the log-likelihood function $\ell(\mathbf{z})$ as we did before, since the likelihood under $\mathcal{H}_1$ depends on an unknown parameter. Instead, we use a simple threshold test on some sufficient statistic $T(\mathbf{z})$.[7]

Consider the detection of some known signal $\mathbf{s}$, with unknown amplitude θ in the presence of noise $\mathbf{n}$ that is Gaussian-distributed with zero mean and covariance matrix $\mathbf{R}$.

$$\mathbf{z} = \theta \mathbf{s} + \mathbf{n} \tag{2.31}$$

First, we determine a sufficient statistic for $\mathbf{z}$ with unknown parameter θ. From [1], we can show that one such sufficient statistic is

$$T(\mathbf{z}) = \sum_{i=1}^{N} \Re\left\{\mathbf{s}^H \mathbf{z}_i\right\} \tag{2.32}$$

for a set of N repeated measurements, $\mathbf{z}_1, \ldots, \mathbf{z}_N$. This represents a matched filter with the known signal $\mathbf{s}$ against each of the measurements $\mathbf{z}_i$, which optimizes the

---

7 This is a result of the Karlin-Rubin Theorem for one-sided composite hypothesis tests, discussed in detail in [1]. There are some underlying assumptions that must be made, most notably that the statistic $T(\mathbf{z})$ must be monotonic in the unknown parameter θ. That is, as θ increases, $T(\mathbf{z})$ must not decrease. If that relatively straightforward requirement is met, then a threshold test on $T(\mathbf{z})$ is a *uniformly most powerful* test for θ, and is said to be optimal in that it maximizes the probability of detection for a given probability of false alarm.

output SNR in the presence of white noise [4]. As the real component of a linear combination of jointly complex Gaussian random variables, the sufficient statistic is Gaussian distributed. For simplicity, we apply a scaling factor that normalizes the variance of $T(\mathbf{z})$ under $\mathcal{H}_0$. If $\mathbf{z}_i = \mathbf{n}$ for all of the samples (the signal is not present), then one can show that the variance of $T(\mathbf{z})$ as computed in (2.32 is

$$E\left[T(\mathbf{z})^2\right] - E\left[T(\mathbf{z})\right]^2 = \frac{NE_s\sigma_n^2}{2} \quad (2.33)$$

where $E_s = \mathbf{s}^H\mathbf{s}$ is the energy in the signal vector. Thus, we introduced the scaled sufficient statistic

$$T(\mathbf{z}) = \sqrt{\frac{2}{NE_s\sigma_n^2}} \sum_{i=1}^{N} \Re\left\{\mathbf{s}^H\mathbf{z}_i\right\} \quad (2.34)$$

Under $\mathcal{H}_0$, this is distributed as a standard normal variable (Gaussian with zero mean and unit variance)

$$T_{\mathcal{H}_0}(\mathbf{z}) \sim \mathcal{N}(0, 1) \quad (2.35)$$

and under $\mathcal{H}_1$, it is distributed again with unit variance, but now with a non-zero mean

$$T_{\mathcal{H}_1}(\mathbf{z}) \sim \mathcal{N}\left(\sqrt{\xi}, 1\right) \quad (2.36)$$

where ξ is an SNR term

$$\xi = \frac{2NE_s\theta^2}{\sigma_n^2} \quad (2.37)$$

The detection threshold is computed from the desired probability of false alarm using the same approach as in (2.26) and (2.27). We begin with the cumulative distribution $F(\mathbf{x})$ in (2.23) to derive the relationship between η and P_{FA}

$$P_{\text{FA}} = 1 - F_{\theta_0}(\eta) \quad (2.38)$$

$$= \frac{1}{2}\left(1 - \text{erf}\left(\frac{\eta}{\sqrt{2}}\right)\right) \quad (2.39)$$

$$\eta = \sqrt{2}\,\text{erfinv}\left(1 - 2P_{\text{FA}}\right) \quad (2.40)$$

$$P_D = \frac{1}{2}\left(1 - \text{erf}\left(\frac{\eta - \sqrt{\xi}}{\sqrt{2}}\right)\right) \quad (2.41)$$

$$= \frac{1}{2}\left(1 - \text{erf}\left(\text{erfinv}\left(1 - 2P_{\text{FA}}\right) - \sqrt{\frac{\xi}{2}}\right)\right) \quad (2.42)$$

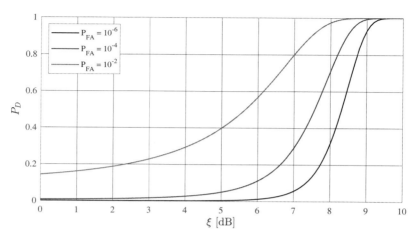

Figure 2.4 Detection performance as a function of the SNR ξ, for various desired probabilities of false alarm. Computed from (2.42) with $\theta = 1$.

This is plotted in Figure 2.4 for various values of P_{FA} and ξ, for the simple case of $\theta = 1$. The MATLAB® code used to generate this plot is the following:

```
% Set up PFA and SNR vectors
pfa = [1e-6,1e-4,1e-2];
xi = 0:.1:10;          % dB Scale
xi_lin = 10.^(xi/10);  % Convert SNR to linear
[PFA,XI] = ndgrid(pfa,xi_lin);

% Compute the PD according to the simplified equation in (2.42)
PD = .5*(1-erf(erfinv(1-2*PFA)-sqrt(XI/2)));
```

2.4 CONSTANT FALSE-ALARM RATE DETECTORS

When the signal to detect is known, but it has unknown amplitude, and the noise has unknown power, then it is impossible to set up a threshold test that guarantees a given P_{FA}, as we did before. The solution is to compute a *constant false-alarm rate* (CFAR) detector. This is a detector that either scales the input signals to account for changes in the noise power or adjusts the detection threshold based on the power of the input noise signal.

From [1], the form of a CFAR detector that optimally detects a known signal (s) in the presence of unknown noise power, is given by a threshold test on the *maximally invariant test statistic*

$$T(\mathbf{z}) = \frac{\mathbf{s}^H \mathbf{P}_s \mathbf{z}}{\sqrt{\mathbf{s}^H \mathbf{s}} \sqrt{\mathbf{z}^H (\mathbf{I} - \mathbf{P}_s) \mathbf{z}}} \qquad (2.43)$$

where $\mathbf{P}_s$ is the projection matrix for the known signal s ($\mathbf{P}_s = \mathbf{s}\mathbf{s}^H / \mathbf{s}^H \mathbf{s}$). The term $\mathbf{s}^H \mathbf{P}_s \mathbf{z}$ serves to compute the amount of received energy that correlates with the desired signal s, while the denominator terms are scale factors that control the size of $T(\mathbf{z})$. The first term ($\sqrt{\mathbf{s}^H \mathbf{s}}$) controls scaling from the multiplication with s, while the second term ($\sqrt{\mathbf{z}^H (\mathbf{I} - \mathbf{P}_s) \mathbf{z}}$) computes the noise power, so that changes in the noise energy do not affect the test statistic $T(\mathbf{z})$.

The test statistic $T(\mathbf{z})$ is distributed according to the *student-t distribution*, discussed in Appendix A [1]. That distribution can be used to compute a threshold η for some desired P_{FA}. Unfortunately, the distribution cannot be inverted in closed form, so numerical methods must be used.

2.4.1 Side Channel Information

Instead of constructing a test statistic that is invariant to changes in the noise level, most systems use *side channel information*, data samples not being used in the current test (because they correspond to other frequencies, or prior time samples that were already tested). *Side channel information* is used to estimate the strength of the noise and interference, and then to adjust the detection threshold accordingly. This is shown graphically in Figure 2.5.

One prominent technique to achieve this is *cell-averaging* (CA)-CFAR. In CA-CFAR, a sliding window is used to compute the average power level of neighboring cells to the *cell under test*. This can be done in any dimension where the noise is expected to be stationary, such as time or frequency. The principal parameters of CA-CFAR are the *window size*, the number of cells averaged to compute an estimate of the noise power, and *guard band size*, the number of cells adjacent to the *cell under test* that are ignored. The guard band is necessary to prevent leakage from a signal in the *cell under tests* from contaminating the neighboring cells and affecting the estimated noise power.

The CA-CFAR estimate is achieved with

$$\bar{\sigma}_n^2 = \frac{1}{K(M-1)} \sum_{i=1}^{M} \bar{\mathbf{z}}_i^H \bar{\mathbf{z}}_i \qquad (2.44)$$

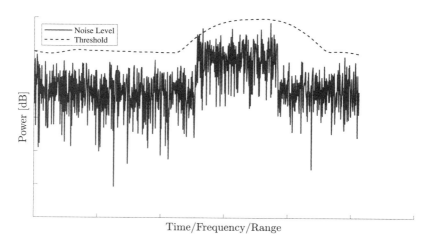

Figure 2.5 Illustration of CA-CFAR via changing threshold levels as the input signal has varying noise power.

where $\bar{\mathbf{z}}_i$, $i = 1, \ldots, M$ are the M side channel sample vectors for the current *cell under test*, and K is the length of the sample vector $\bar{\mathbf{z}}_i$. The output $\bar{\sigma}_n^2$ is an estimate of the noise power σ_n^2. We divide the noise power estimate by M-1 instead of by M to ensure that it is an *unbiased* estimate. (This is discussed in more detail in Chapter 6.) We reuse the test statistics from (2.32) and the detection threshold from (2.40), replacing the actual covariance matrix σ_n^2 with the estimated value $\bar{\sigma}_n^2$

$$T(\mathbf{z}) = \frac{1}{\sqrt{2NE_s}} \sum_{i=1}^{K} \Re\left\{\mathbf{s}^H \mathbf{z}_i\right\} \qquad (2.45)$$

$$\eta = \sqrt{2\bar{\sigma}_n^2} \operatorname{erfinv}\left(1 - 2P_{\text{FA}}\right) \qquad (2.46)$$

Note that the test statistic $T(\mathbf{z})$ is no longer scaled by the noise power σ_n^2. This scale factor is now accounted for by including the noise power estimate $\bar{\sigma}_n^2$ in the variable threshold η.

2.5 PROBLEM SET

2.1 For Gaussian random trials, the statistic

$$\bar{x} = \sum_{m=0}^{M-1} x_m$$

is a sufficient statistic for the mean μ of the samples x_m. Compare the number of bits required to encode/store/transmit $\bar{x}$ to that required for the entire sample vector $\mathbf{x} = [x_0, x_1, \ldots, x_{M-1}]$.

2.2 Consider a measurement vector distributed as a Rayleigh random variable $\mathbf{z} \sim \mathcal{R}(\sigma)$. Prove that $t = \sum_{n=0}^{N-1} x_i^2$ is a sufficient statistic for σ^2.

2.3 Consider the binary hypothesis test $\mathbf{z} \sim \mathcal{CN}(\boldsymbol{\mu}, \theta \mathbf{I})$ for an M-element data vector.

$$\begin{aligned} \mathbb{H}_0 &: \theta = \theta_0 \\ \mathbb{H}_1 &: \theta = \theta_1 \end{aligned}$$

Derive the log-likelihood ratio test, assuming that $\mathbf{m}$, θ_0 and θ_1 are known.

2.4 Consider a set of N-dimensional sample vectors that are independent and identically distributed with $\mathbf{x}_m \sim \mathcal{N}(\mathbf{H}\boldsymbol{\vartheta}, \mathbf{C})$ and M snapshots collected to form the data matrix $\mathbf{X} = [\mathbf{x}_0, \ldots, \mathbf{x}_{M-1}]$. Find a sufficient statistic $\mathbf{T}(\mathbf{x})$ for $\boldsymbol{\vartheta}$. What is its variance?

2.5 Assume a Gaussian test with $x \sim \mathcal{CN}\left(\theta s e^{j\phi}, \sigma^2 \mathbf{I}\right)$, $\theta = 0$ under $\mathcal{H}_0$, $\theta > 0$ under $\mathcal{H}_1$, σ^2 known, and $\mathbf{s}$ and ϕ unknown. The optimal detector, in this case, is a scaled energy detector $z = \mathbf{x}^H \mathbf{x}/\sigma^2$. Let $\xi = \mathbf{s}^H \mathbf{s}/N\sigma^2 = 1$. What is the distribution of z under both $\mathcal{H}_0$ and $\mathcal{H}_1$? Plot the threshold η as a function of $P_{\text{FA}} \in [10^{-9}, 10^{-1}]$, for MN=10, 100, and 200. Plot P_D as a function of P_{FA}.

2.6 Consider a binary hypothesis test with

$$\begin{aligned} f_X(x|\mathcal{H}_0) &= \alpha e^{-x/\alpha} & x \geq 0 \\ f_X(x|\mathcal{H}_1) &= \beta e^{-x/\beta} & x \geq 0 \end{aligned}$$

where $\beta > \alpha$. What is the log likelihood ratio? Simplify it to remove constants. Find the expression for η as a function of P_{FA}, and for P_D as a function of η and the scale constants α and β. Plot P_D as a function of $P_{\text{FA}} \in [10^{-6}, 10^{-1}]$ for $\alpha/\beta = .1, .5,$ and $.9$.

2.7 Consider the detection of a constant $m = 2$ in complex Gaussian noise with variance $\sigma_n^2 = 4$. Let the number of samples be $N = 10$. What is the detector SNR d^2? Suppose $P_{\text{FA}} = 0.01$, what is the value of the threshold η? What is the achievable P_{D}?

References

[1] L. L. Scharf, *Statistical signal processing*.   Reading, MA: Addison-Wesley, 1991.

[2] H. L. Van Trees, *Detection, Estimation, and Linear Modulation: Part I of Detection, Estimation, and Modulation Theory*.   Hoboken, NJ: Wiley, 2001.

[3] S. M. Kay, *Fundamentals of Statistical Signal Processing, Volume 2: Detection Theory*.   Upper Saddle River, NJ: Prentice Hall, 1998.

[4] M. A. Richards, *Fundamentals of Radar Signal Processing*.   New York, NY: McGraw-Hill Education, 2014.

Chapter 3

Detection of CW Signals

This chapter considers the detection of a CW signal, with some finite bandwidth. Subsequent chapters will focus on detection of spread spectrum signals and of signals with an unknown carrier frequency (via scanning or searching for the channel in use).

This chapter relies heavily on statistical signal processing theory in [1]. Similar detection performance derivations can be found in [2], with a focus on radar processing in [3], and with a focus on EW detection of communications signals in [4].

We begin with a background on CW signal detection, formulate the problem in terms familiar from Chapter 2, and provide the solution. We then discuss some real-world examples of CW signal detection.

3.1 BACKGROUND

The simplest version of a CW signal is a tone, but this class of signals can also include modulated signals, such as frequency modulation (FM), phase modulation (PM), or amplitude modulation (AM). In this chapter, we do not make any assumptions on the signal modulation. Rather, the defining characteristics of a CW signal are the infinite temporal support (it is always on, as opposed to *pulsed* signals) and finite spectral support (spectral content is limited to some band).

The parameters of the signal to be detected, such as the center frequency (f_0), bandwidth (B_s), and modulation type, are assumed to be unknown. However, to make the problem tractable, we assume that some frequency channel is being

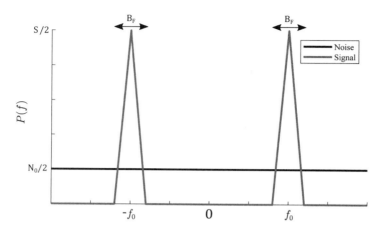

Figure 3.1 Power spectral density (PSD) of noise and a band-limited signal.

monitored, and that the CW signal (if it exists) is in that channel. The alternative case is discussed in detail in Chapter 5.

3.2 FORMULATION

Consider a signal $s(t)$ with central frequency f_0 and bandwidth B_s, illustrated by Figure 3.1. The exact power spectral density of the signal within that band is unknown (the triangular shape is notional). The noise, $n(t)$, is assumed to be additive, white Gaussian noise (AWGN).[1] The noise density is N_0 (watts/hertz), and is shown in Figure 3.1. Note that the amplitude of the noise in Figure 3.1 is $N_0/2$; this is because the noise energy is equally split between the positive frequencies and quadrature.

The first step in detection is to pass the incoming signals through a bandpass filter, with bandwidth B_F. This serves to suppress any unwanted signals that are out-of-band, and to reduce the total amount of energy that enters the detector. The effect of this bandpass filter is shown in Figure 3.2. The total noise power leaving

---

[1] "White" in AWGN refers to a flat power spectral density, and "Gaussian" refers to the statistical distribution of each sample in time. Both the in-phase and quadrature components of the noise are independent and Gaussian-distributed.

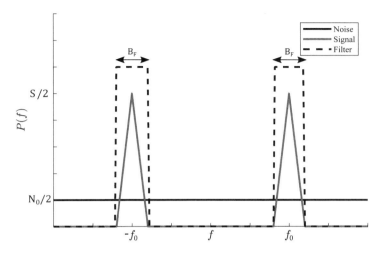

Figure 3.2 PSD of receive signal, noise, and bandpass filter response.

the bandpass filter is computed

$$N = \frac{N_0}{2} \int_{-\infty}^{\infty} |H(f)|^2 \, df \tag{3.1}$$

where $H(f)$ is the bandpass filter response. If we assume an ideal filter with response

$$H(f) = \begin{cases} 1 & |f - f_0| \leq \frac{B_F}{2} \\ 1 & |f + f_0| \leq \frac{B_F}{2} \\ 0 & \text{otherwise} \end{cases} \tag{3.2}$$

then the noise power simplifies to the product of the noise power density (both I and Q) and the passband bandwidth

$$N = N_0 B_F \tag{3.3}$$

It is clear that the passband of the filter must be as narrow as possible, to reduce the amount of noise entering the receiver. However, if it is smaller than the bandwidth of the signal B_s, then it will also reduce the amount of signal entering the filter. Define the signal's PSD $P_s(f)$; the signal power entering the filter is

$$S = \int_{-\infty}^{\infty} P_s(f) \, df \tag{3.4}$$

and leaving the filter is

$$\widetilde{S} = \int_{-\infty}^{\infty} P_s(f) \, |H(f)|^2 \, df \tag{3.5}$$

If we again assume the ideal filter response in (3.2), and make a similar assumption on the signal PSD

$$P_s(f) = \begin{cases} 1 & |f - f_0| \leq B_s \\ 1 & |f + f_0| \leq B_s \\ 0 & \text{otherwise} \end{cases} \tag{3.6}$$

then the output signal power is described by the simplified expression

$$\widetilde{S} = S \min\left[1, \frac{B_F}{B_s}\right]. \tag{3.7}$$

For simplicity, we assume that $B_F > B_s$, and that $\widetilde{S} \approx S$.

To define the detection problem, we convert the signal to baseband, and digitize the I and Q samples of the incoming signal, sending M complex samples to the detector. The signal is written

$$\mathbf{z} = \mathbf{s} + \mathbf{n} \tag{3.8}$$

under $\mathcal{H}_1$. The noise vector is complex Gaussian with variance σ_n^2 as defined in (3.3)

$$\mathbf{n} \sim \mathcal{CN}\left(0, \sigma_n^2 \mathbf{I}_M\right) \tag{3.9}$$

where σ_n^2 is the variance (or power) in a single sample, and $\mathbf{I}_M$ is an $M \times M$ identity matrix. Given a sampling rate of F_s Hz, the noise sample variance is computed from the power entering the receiver

$$\sigma_n^2 = \frac{N}{F_s} \tag{3.10}$$

An important caveat to sampling the noise is that if F_s is greater than the Nyquist sampling rate, the noise samples will no longer be independent (as a result of the Nyquist sampling theorem). The signal (whose structure is assumed unknown) has total energy given by

$$E_s = ST = \frac{SM}{F_s} = \mathbf{s}^H \mathbf{s} \tag{3.11}$$

where S is the signal power (in watts) coming out of the bandpass filter, and T is the observation interval (in seconds). The resulting energy E_s is expressed in joules. We

Detection of CW Signals

show an alternative expression, where we see the relationship between the sampling rate and the number of samples (note that the number of samples M is computed $M = \lfloor TF_s \rfloor$). Finally, this can also be expressed as the magnitude squared of the sample vector **s**.

From this, we express the binary hypothesis

$$\begin{array}{ll} \mathcal{H}_0: & \mathbf{z} = \mathbf{n} \quad \sim \mathcal{CN}\left(0, \sigma_n^2 \mathbf{I}_M\right) \\ \mathcal{H}_1: & \mathbf{z} = \mathbf{s} + \mathbf{n} \quad \sim \mathcal{CN}\left(\mathbf{s}, \sigma_n^2 \mathbf{I}_M\right) \end{array} \quad (3.12)$$

3.3 SOLUTION

If the signal **s** were known, then the solution to (3.12) would be a threshold test on the matched filter, as described in (2.34). However, since **s** is not known, we must find a test statistic that does not rely on it. In this case, the best that can be done without making any assumptions on the structure of **s** is an energy detector[2]

$$T(\mathbf{z}) = \mathbf{z}^H \mathbf{z} \quad (3.13)$$

If we can assume that the noise power is known, or perfectly estimated, then it is straightforward to scale the test statistic

$$T(\mathbf{z}) = \frac{2}{\sigma_n^2} \mathbf{z}^H \mathbf{z} \quad (3.14)$$

The scale factor under the null hypothesis is the component noise power in the I and Q dimensions, which is defined as half the noise power for white Gaussian noise signals. This scale has the effect of making the sufficient statistic $T(\mathbf{z})$ follow a chi-squared distribution with 2M degrees of freedom (see Appendix A), written

$$T_{\mathcal{H}_0}(\mathbf{z}) \sim \chi_{2M}^2 \quad (3.15)$$

The PDF of a chi-squared random variable with ν degrees of freedom is given by

$$f_{\chi^2}(x|\nu) = \frac{1}{2^{\nu/2}\Gamma(\nu/2)} x^{\nu/2-1} e^{-x/2} \quad (3.16)$$

---

2 If anything about **s** is known, such as the modulation scheme, or a subspace in which **s** must reside, then projection of **z** onto that subspace can reduce the amount of noise present in the detector. See [1] for derivation of subspace energy detectors.

where $\Gamma(\nu/2)$ is the gamma function.[3]

Under $\mathcal{H}_1$, the components of $\mathbf{z}$ have nonzero mean. The result is that $T(\mathbf{z})$ is a *non-central* chi-squared random variable, again with 2M degrees of freedom, but now with noncentrality parameter

$$\lambda = \frac{E_s}{\sigma_n^2/2} = \frac{MS}{\sigma_n^2/2} = \frac{2MS}{N} \tag{3.17}$$

We write this distribution

$$T_{\mathcal{H}_1}(\mathbf{z}) \sim \chi_{2M}^2(\lambda) \tag{3.18}$$

The PDF of a noncentral chi-squared random variable with ν degrees of freedom and noncentrality parameter λ is given by

$$f_{\chi^2}(x|\nu,\lambda) = \sum_{i=0}^{\infty} \frac{e^{-\lambda/2}(\lambda/2)^i}{i!} f_{\chi^2}(x;\nu+2i) \tag{3.19}$$

For simplicity, we define the SNR ξ as

$$\xi = \frac{S}{N} \tag{3.20}$$

Thus, we can rewrite the distribution

$$T_{\mathcal{H}_1}(\mathbf{z}) \sim \chi_{2M}^2(2M\xi) \tag{3.21}$$

3.3.1 Threshold Selection

As in Chapter 2, we determine the threshold by fixing the desired P_{FA}, and solving the equation

$$P_{\text{FA}} = P_{\mathcal{H}_0}\{T(\mathbf{z}) \geq \eta\} \tag{3.22}$$

Using the PDF of $T(\mathbf{z})$ under $\mathcal{H}_0$, we express the P_{FA} in integral terms, and simplify, using the cumulative distribution function of a chi-squared random variable $\left(F_{\chi^2}(x;\nu)\right)$

$$P_{\text{FA}} = \int_{\eta}^{\infty} f_{\chi^2}(u;2M)\,du \tag{3.23}$$

$$= 1 - \int_{-\infty}^{\eta} f_{\chi^2}(u;2M)\,du \tag{3.24}$$

$$= 1 - F_{\chi^2}(\eta;2M) \tag{3.25}$$

---

[3] The gamma function is $\Gamma(z) = \int_0^{\infty} x^{z-1}e^{-x}dx$. For positive integers n, it can be simplified with the closed-form expression: $\Gamma(n) = (n-1)!$.

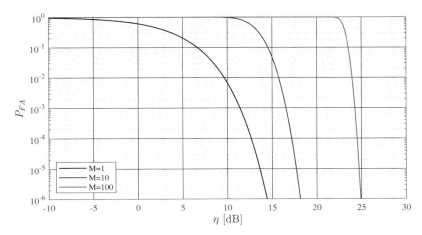

Figure 3.3 Probability of false alarm for the CW signal detector in (3.15), calculated according to (3.25).

There is no closed-form solution, but many programming languages have the noncentral chi-squared distribution included. To compute P_{FA} in MATLAB® for a given threshold, use the chi2cdf function[4]

```
Pfa = 1-chi2cdf(eta,2*M);
```

Similarly, to compute a threshold η from a desired P_{FA}, use the function chi2inv

```
eta = chi2inv(1-Pfa,2*M);
```

This has no closed-form solution, but P_{FA} is plotted in Figure 3.3 for several values of N as a function of η, using the code snippets presented above.

3.3.2 Detection Algorithm

The detector presented is a simple threshold test of the sufficient statistic $T(\mathbf{z})$ against η. This detector is commonly referred to as a *square-law* detector, since it uses the square of each sample in the input vector $\mathbf{z}$. This is in contrast to

---

[4] The chi2cdf and chi2inv MATLAB® functions are part of the *Statistics & Machine Learning Toolbox*.

a linear detector, which uses the amplitude of each sample. The implementation of this detector is included in the provided MATLAB® code, in the function `detector.squareLaw`.

3.3.3 Detection Performance

Similarly, from the computed threshold, the achieved P_D (for a given σ_s^2) is computed by taking the inverse of the cumulative distribution of the noncentral chi-squared random variable described in (3.21) for the threshold η

$$P_D = 1 - F_{\chi^2}(\eta; 2M, 2M\xi) \qquad (3.26)$$

This does not have a closed-form solution[5] but can be computed with the MATLAB® function `ncx2cdf`[6]

```
Pd = 1 - ncx2cdf(eta,2*M,2*M*xi));
```

The probability of detection according to this equation is plotted as a function of ξ in Figure 3.4, for several values of M and desired P_{FA} values. The MATLAB® code used to generate this plot is the following.

```
% Initialize parameters
PFA = 1e-6;
M = [1 10 100];
xi_db = -10:.1:20;
xi_lin = 10.^(xi_db/10);
[MM,XI] = ndgrid(M,xi_lin);

% Compute threshold and Probability of Detection
eta = chi2inv(1-PFA,2*MM);
PD = 1 - ncx2cdf(eta,2*MM,2*MM.*XI);
```

3.4 PERFORMANCE ANALYSIS

Now that we have defined the necessary detector for CW signals, we will discuss two example scenarios, detection of a (very powerful) FM radio transmission, as well as detection of an airborne collision avoidance radar.

---

[5] The cumulative distribution of a noncentral chi-squared random variable can be written in terms of the Marcum-Q function, which does not have a closed-form representation [5].

[6] The `ncx2cdf` MATLAB® function is included in the *Statistics & Machine Learning Toolbox*.

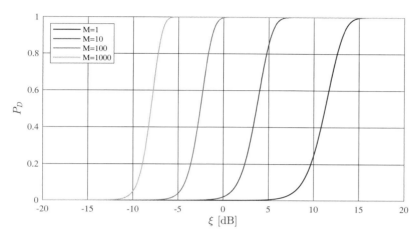

Figure 3.4 Probability of detection for the energy detector of (3.15) as a function of SNR ξ, for $P_{FA} = 10^{-6}$ [calculated according to (3.26)].

3.4.1 Brief Review of RF Propagation

Before we present some example scenarios, it is necessary to briefly introduce some principles of RF propagation. These concepts are discussed in more detail in Appendix B. In this section, we review three principles: free-space path loss, two-ray path loss, and atmospheric absorption. The one-way link equation that relates transmitted and received power is given by[7]

$$S = \frac{P_t G_t G_r}{L_{\text{prop}} L_{\text{atm}} L_t L_r} \quad (3.27)$$

where P_t is the transmit power, G_t is the gain of the transmit antenna, G_r is the gain of the receive antenna, L_{prop} is the propagation loss (such as free space or two-ray), L_{atm} is the atmospheric loss, L_t encompasses all system losses in the transmitter to be detected, and L_r encompasses all losses in the detector hardware and processing. Sections 3.4.1.1 and 3.4.1.2 introduce formulas for L_{prop}, and section 3.4.1.3 discusses L_{atm}. Note that this is a simple equation and can be supplemented with additional specific system losses, such as filter losses and amplifier losses, or those can be rolled into the broader L_t and L_r terms.

---

[7] Received power is written P_r in many texts; we use the notation S to be consistent with the formulation of the optimal detector earlier in this chapter.

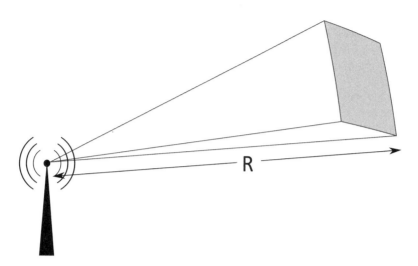

Figure 3.5 Illustration of free-space path loss. The spreading is described by the surface area of a sphere with radius equal to the propagation distance.

3.4.1.1 Free-space Path Loss

When an RF signal spreads unimpeded, the energy density along the wavefront scales with the inverse of the propagation distance squared; see Figure 3.5. The path loss, which describes the ratio of the received power (by an isotropic antenna) to the emitter power (in the direction of the receiver), is given by

$$L_{\text{fspl}} = \left(\frac{4\pi R}{\lambda}\right)^2 \tag{3.28}$$

Free-space path loss is a valid assumption when there are no obstructions between the transmitter and receiver, and there are no dominant reflectors, such as a building or the ground, near the line of sight between the transmitter and receiver.

3.4.1.2 Two-Ray Path Loss

In the presence of a dominant reflector (such as the Earth's surface), the free-space attenuation model is no longer adequate, as the dominant reflection begins to overlap with the direct path signal in time, causing destructive interference.

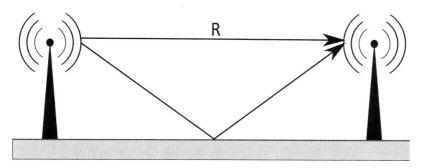

Figure 3.6 In the two-ray model, reflections from the Earth's surface are the primary source of destructive interference.

See Figure 3.6 for an illustration of this ground bounce path. The received signal strength was found empirically to vary with the fourth power of range, and inversely with the square of the transmitter and receiver heights. Interestingly, this path loss model is independent of wavelength

$$L_{2-\text{ray}} = \frac{R^4}{h_t^2 h_r^2} \qquad (3.29)$$

where h_t and h_r are the height of the transmitter and receiver above the Earth, respectively.

For any terrestrial link, there is a range beyond which the free-space model is no longer appropriate and two-ray should be used. This is referred to as the *Fresnel zone*, and is computed [6]

$$\text{FZ} = \frac{4\pi h_t h_r}{\lambda} \qquad (3.30)$$

Thus, we have a simple piecewise propagation model

$$L_{\text{prop}} = \begin{cases} \left(\frac{4\pi R}{\lambda}\right)^2 & R \leq \frac{4\pi h_t h_r}{\lambda} \\ \frac{R^4}{h_t^2 h_r^2} & \text{otherwise} \end{cases} \qquad (3.31)$$

An implementataion of (3.31) is provided in the MATLAB® code that accompanies this text, and can be called with the code snippet as follows.[8]

---

8 MATLAB® provides a number of channel models as part of the *Phased Array Systems Toolbox*, but these are geared toward modeling actual transmissions, and calculation of path loss is not explicitly returned.

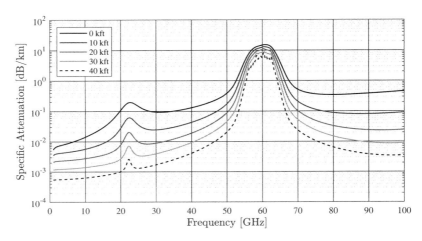

Figure 3.7 Atmospheric attenuation for standard atmospheric assumptions at a series of altitudes (includes both dry air and water vapor, but not rain or fog).

```
> prop.pathLoss(R,f0,ht,hr,includeAtmLoss);
```

Please see Appendix B for additional details, including more complex propagation models for specific situations.

3.4.1.3 Atmospheric Absorption

In addition to attenuation of the signal energy as it spreads and propagates through the atmosphere, there is an absorption effect, caused by the electric and magnetic dipole effects of gases in the atmosphere. Figure 3.7 shows specific attenuation as a loss of α dB for every kilometer of propagation distance, across a series of altitudes.[9] The impact of atmospheric attenuation is heavily dependent on temperature, pressure, and vapor density. This results in the loss equation

$$L_{\text{atm}}[dB] = \alpha \frac{R}{1e3} \tag{3.32}$$

Note that (3.32) is expressed natively in log format, while the others are presented linearly. This is an important distinction to remember when calculating atmospheric

---

9   See Appendix C for details on atmospheric assumptions used to generate this chart.

Detection of CW Signals 41

loss. More detail, including a table of atmospheric loss values across frequencies, and models for absorption by rain and clouds, is included in Appendix C.

Absorption by atmospheric gases is often ignored for frequencies below 1 GHz [7].

3.4.2 Detection of an FM Broadcast Tower

Some of the most proliferated, and easy-to-detect, CW transmission signals are television and radio broadcasts, which are transmitted at very high power, from high atop a tower to minimize ground effects on propagation. In the United States, the FM radio band consists of 100 channels, each 200 kHz wide, between 88 MHz and 108 MHz.

The first step is to compute the received power level, as a function of range. Since the carrier frequency of the FM transmission is below 1 GHz, we can ignore atmospheric attenuation.

We turn now to the transmission source. Engineers and analysts frequently choose to replace $P_t G_t / L_t$ with the simplified term *effective radiated power* (ERP). A typical ERP is 47 dBW (50 kW) or less for FM broadcast towers. Thus, for a given range, the received power is 47 dBW minus the total receiver and path losses. This is plotted in Figure 3.8.

Next, we must compute the noise power in the receiver. Consider a matched receiver with a 200-kHz passband that aligns with the transmit channel (the full signal power is received). Noise calculations are covered in Appendix D, but let us assume a 5-dB noise figure (F_N). This results in a noise power density into the receiver of[10]

$$N_0 = kT10^{F_n/10} = 1.266 * 10^{-20} \text{W/Hz} \quad (3.33)$$
$$N_0[dB] = 10 log_{10}(N_0) = -198.97 \text{dBW/Hz} \quad (3.34)$$

We multiply this noise power density by the passband bandwidth of 200 kHz to compute the total noise power

$$N = N_0 B_f = 2.53 * 10^{-15} \text{W/Hz} \quad (3.35)$$
$$N[dB] = 10 \log_{10}(N_0 B_f) = -145.97 \text{dBW} \quad (3.36)$$

---

10   k is Boltzmann's constant ($1.38*10^{-23} m^2 Hz^2/kg*K$), and T_0 is the standard noise temperature (290K). See Appendix D for details.

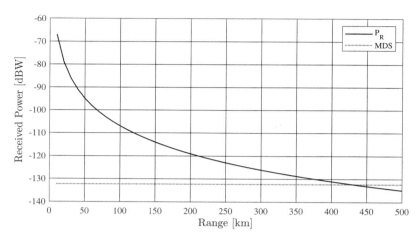

Figure 3.8 Received power (solid line) and minimum detectable signal (dotted line) as calculated in (3.38) with parameters from Section 3.4.2.

We now compute the receiver SNR with the equation

$$\xi = \frac{S}{N} = \frac{P_t G_t G_r}{L_{\text{prop}} L_{\text{atm}} L_t L_r N} \qquad (3.37)$$

Example 3.1 Maximum Detection Range of an FM Radio Station

Given the detector and transmitter described above, what is the maximum range at which the detector will satisfy a false alarm rate of $P_{\text{FA}} = 10^{-6}$ and detection rate of $P_{\text{D}} = 0.5$ with $M = 10$ samples?[11]

Ideally, we would like a closed-form equation to compute the required SNR ξ given these parameters. However, there is no closed-form solution to inverting (3.26) and solving for ξ. For the specified probabilities of detection and false alarm, and the number of samples, one can inspect Figure 3.4 and determine that $\xi \approx 3.5$ dB satisfies these constraints. To find the exact answer, we test a range of SNR values around that result.

---

[11] Assuming a matched detector with 200-kHz bandwidth, we must sample the received signal at the Nyquist rate of 400 kHz (twice the bandwidth) for real valued samples or 200 kHz for complex-valued samples. Thus, the $M = 10$ samples represent an integration time of 10/400e3=25 μs.

Detection of CW Signals

The provided MATLAB® code contains a script designed to search for the true minimum SNR to achieve a desired operating point, under the function `detector.squareLawMinSNR`.

The result of this code is 3.65 dB. We can compute the minimum detectable signal using this required SNR, along with the noise level from (3.36).

$$MDS = \xi_{\min}[dB] + N[dB] - G_r + L_r = -142.32 \text{dBW} \quad (3.38)$$

where G_r is the receiver gain and L_r represents the system losses; both here are assumed to be zero, for simplicity. The MDS represents the weakest signal power (as received by an omnidirectional antenna) that a receiver can detect, and it takes into account any antenna gain or system losses.

At this point, one could consult Figure 3.8 to find the range at which $\xi \geq 3.65$ dB, or when $P_r \geq -142.32$ dBW. The result is roughly 430 km. For an exact answer, we can employ an iterative optimization algorithm. One such algorithm is included in the provided MATLAB® code, under the function `detector.squareLawMaxRange`.

$$\xi_0 = \frac{P_t G_t G_r}{L_t L_r N} \quad (3.39)$$

and the `includeAtmLoss` flag dictates whether atmospheric absorption is to be considered (using the accompanying `atmStruct` definition, if needed).[12] The result of this code is 429.88 km.

From this calculation, it is possible to detect an FM radio transmission at very long distances. However, given the curvature of the Earth, the signal is likely to be blocked by terrain. The radio horizon at 100m for a receiver at 2m is roughly 40 km, indicating that the signal will be blocked by terrain far sooner than the signal will be lost below the noise floor. See Appendix B for further discussion of the radio horizon.

As a verification of this calculation, we can run a Monte Carlo simulation. Figure 3.9 plots results for 10^5 trials at each standoff distance between 100 km and 1,000 km. The code for this trial is provided on the provided MATLAB® package and makes use of the square law detector function introduced earlier in this chapter. The MATLAB® code to run this trial is included in the `examples` folder of the provided MATLAB® code.

---

12 The atmospheric terms are discussed in Appendix C. The standard reference atmosphere can be loaded with a call to `prop.makeRefAtmosphere()`;

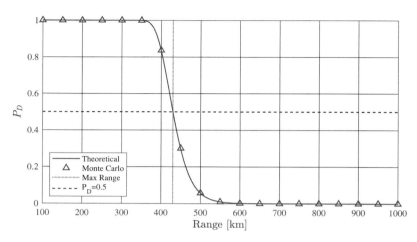

Figure 3.9 Monte Carlo trial plotting probability of detection for an FM tower at various standoff ranges from an ideal receiver, with 10^5 trials at each range.

3.4.3 CW Radar Detection

Next, consider detection of a continuous wave radar, such as that defined in Section 2.4 of [8]. The radar in question is a collision avoidance radar built by NASA Ames and Honeywell; key parameters are summarized in Table 3.1.[13] This particular radar is coded with a binary signal onto the phase of the carrier signal.

Example 3.2 Maximum Detection Range of a CW Radar

We wish to detect the presence of this radar aboard a helicopter, with our detector mounted on another helicopter. Detection of the collision avoidance radar is used as an early warning detection of the approaching threat platform. We assume that both helicopters are airborne at an altitude of $h = 500$m.

The receiver has a matched 40-MHz bandwidth and a 4-dB noise figure, resulting in a noise level of -124 dBW. If we assume the same detector requirements as in the previous example ($P_D = .5$, $P_{FA} = 10^{-6}$, $M = 10$), then the SNR requirement is unchanged (3.66 dB), and we can compute the minimum detectable

---

13 See Section 2.4 of [8] for additional details.

Table 3.1
Collision Avoidance Radar Parameters for Example 3.2

(a) CA Radar Transmitter

Parameter	Value
P_t	35 mW
G_t (mainlobe)	34 dBi
G_t (sidelobe)	19 dBi
B_s	31.3 MHz
f_0	35 GHz
L_t	0 dB
h_t	500m

(b) Receiver

Parameter	Value
G_r	10 dBi
L_r	2 dB
F_n	4 dB
B_n	40 MHz
h_r	500m

signals.[14] In this case, let us assume a receive gain of 10 dBi, representing a moderately directional receiver, and receiver losses of 2 dB.

$$MDS = \xi_{\min}[dB] + N[dB] - G_r + L_r = -128.3 \text{dBW} \quad (3.40)$$

Next, we compute the receiver power level and compare to this threshold for mainlobe ($G_t = 34$ dBi) and sidelobe ($G_t = 34 - 25 = 9$ dBi) scenarios. Note that, given the height of the transmitter and receiver, free-space path loss is in effect.[15] The atmospheric loss at 35 GHz occurs at a rate of 0.11 dB/km. Mainlobe detection occurs at a range of 13.9 km, while sidelobe detection occurs at a range of 923m. Compare these numbers to the radar's operational range of 900m [8]. Based on these numbers, it is possible to detect the mainlobe of the radar's emissions far before it can detect the ownship's presence (10.3 km versus 900m), but detection of the radar's sidelobe pulses will occur shortly before the radar has a chance to detect the EW receiver's platform.[16]

Figure 3.10 plots the resutls of a Monte Carlo verification of the predicted detection range for mainlobe and sidelobe detection. The experiment confirms that mainlobe signals from the radar can be detected at a range of 13.9 km, given the

---
14 Note that, assuming we sample the signal at the Nyquist rate (twice the signal bandwidth), 10 independent detection samples represent an integration time of $10F_s = .16\mu$ s.
15 The *Fresnel zone* is more than 366,000 km.
16 The radio horizon for these transmitters is 160 km, so it does not affect detection range in this example.

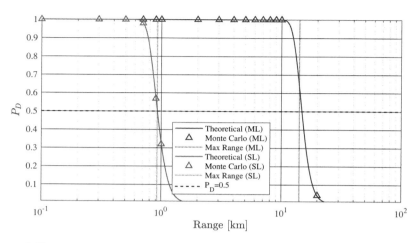

Figure 3.10 Monte Carlo trial results plotting probability of detection for Example 3.2 at various ranges, with 10^5 trials at each standoff range.

specified receiver settings. Code to run this experiment is included in the provided MATLAB® package, under the examples folder.

3.5 PROBLEM SET

3.1 Consider the binary hypothesis test in (3.12) and the energy detector in (3.14). What is the threshold η if we desire $P_{FA} = .01$? What is $P_{FA} = 10^{-3}$? What is $P_{FA} = 10^{-6}$? Assume $M = 10$.

3.2 In Problem 3.1, what is the required SNR ξ to achieve $P_D = 0.5$ at each of the desired P_{FA} levels?

3.3 What is the free-space propagation loss at 100 km for a 10-GHz signal?

3.4 Consider a transmitter and receiver at the same height ($h_r = h_t = h$). What is the minimum height h such that a 10-GHz transmission has a Fresnel zone distance of 100 km? What is the path loss using the two-ray model?

3.5 Consider a transmitter and receiver at a height of $h = 10$ kft (10,000 feet). Using Figure 3.7, determine the atmospheric loss due to gases and water vapor for $f = 1$ GHz, 10 GHz, and 30 GHz. What is the total loss for each frequency at a range of 50 km?

3.6 Automotive radars frequently use the 74–76-GHz band. Assuming an ERP of 20 dBW, a carrier of 75 GHz, and $h_t = 1$m, what is the loss for a receiver at $h_r = 10$m at a range of 1 km? At what range will the radar be detected by a receiver with a minimum detectable signal of -120 dBW?

3.7 Consider Example 3.2. If the integration period is set to $T = 1$ ms, what is the number of samples M (assuming Nyquist sampling)? What is the resulting maximum detection range (for both mainlobe and sidelobe transmissions)?

References

[1] L. L. Scharf, *Statistical signal processing.* Reading, MA: Addison-Wesley, 1991.

[2] H. L. Van Trees, *Detection, Estimation, and Linear Modulation: Part I of Detection, Estimation, and Modulation Theory.* Hoboken, NJ: Wiley, 2001.

[3] M. A. Richards, *Fundamentals of Radar Signal Processing.* New York, NY: McGraw-Hill Education, 2014.

[4] R. Poisel, *Modern Communications Jamming: Principles and Techniques, 2nd Edition.* Norwood, MA: Artech House, 2011.

[5] M. K. Simon, *Probability distributions involving Gaussian random variables: A handbook for engineers and scientists.* New York, NY: Springer Science & Business Media, 2007.

[6] D. Adamy, *EW 103: Tactical battlefield communications electronic warfare.* Norwood, MA: Artech House, 2008.

[7] ITU-R, "Rec: P.676-11, Attenuation by atmospheric gases in the frequency range 1–350 GHz," International Telecommunications Union, Tech. Rep., 2016.

[8] M. A. Richards, J. Scheer, W. A. Holm, and W. L. Melvin, *Principles of Modern Radar.* Edison, NJ: SciTech Publishing, 2010.

Chapter 4

Detection of Spread Spectrum Signals

For as long as the electromagnetic spectrum has been used to communicate or to sense, it has been used to detect adversaries by their emissions. In response to this, a class of protected signals are widely employed. These signals have varying levels of protection and are broadly lumped into three categories: *low probability of detection* (LPD), *low probability of interception* (LPI), and *low probability of exploitation* (LPE). A signal that is LPD is, by definition, also LPI and LPE. Similarly, one that is LPI is also LPE.

The first, LPD, refers to signals that are difficult to detect. The most popular way to achieve LPD is to spread signal energy over a wide bandwidth, such as with DSSS communications signals. By diluting the energy over a wide bandwidth, there is less likelihood that a receiver will successfully detect the presence of the signal.

The second, LPI, refers to protection from an adversary intercepting individual pieces of a transmission. The adversary may be able to detect that a transmission is occurring but is unable to intercept the entire signal. The principal benefit of LPI is that the adversary has a harder time estimating signal parameters, since they capture smaller portions of the signal. A popular way to achieve LPI is through frequency hopping, which is discussed in detail in Chapter 5.

The final class of protected signal is LPE, which refers to protection from an adversary exploiting signal characteristics, such as the information transmitted. The principal method of achieving LPE is through encryption.

This chapter discusses spread spectrum signals, of which DSSS is a type, and a number of specialized detectors that are well suited to these challenging signals.

4.1 BACKGROUND

DSSS refers to the use of a much larger portion of the EMS to perform a transmission than is required. By *spreading* a transmitted signal across bandwidths in excess of 10–100x its original bandwidth, the transmitted energy is diluted significantly, often to the point where the PSD of the transmitted signal is lower than the background noise level, thereby making detection via traditional means (energy detectors) much more challenging.

Frequency-hopping spread spectrum (FHSS) is a related technique for spreading a signal over much wider bandwidth, but it relies on rapid changes in center frequency and at any given instant is a narrowband signal. It is discussed in Chapter 5.

DSSS is a increasingly used in both military and civilian contexts. In addition to its military benefits, DSSS can reduce interference and provide a way for multiple users to occupy a single channel.

Radar signals often exhibit similar characteristics to DSSS communications signals, most notably phase shift keying (PSK) and noise radar waveforms exhibit the same spread spectral behavior as DSSS signals [1], in that they occupy a wide instantaneous bandwidth. For radar signals, a large bandwidth is often advantageous (since resolution is inversely proportional to bandwidth), but it can complicate the processing required to detect large targets that span multiple resolution cells [2].

In either case, the defining structure of digital spread spectrum signals is that they consist of a series of repeated "chips." How these chips vary is a function of the modulation scheme chosen, such as binary or polyphase PSK, or quadrature amplitude modulation. The time that each chip is transmitted is referred to as the chip duration: T_{chip}. An example of this is shown in Figure 4.1, which depicts the transmission of a 5-bit sequence (01101) using BPSK modulation. The frequency spectrum is generated by mixing this signal with a carrier at roughly 4x the frequency.[1] The carrier frequency offset, and spectral bandwidth, are shown.

The bandwidth of the signal is the inverse of the chip rate (barring any windowing functions used to smooth the spectral shape)

$$B_s = \frac{1}{T_{\text{chip}}} \tag{4.1}$$

The relationship between chip rate and bandwidth is shown in Figure 4.2.

---

[1] To generate the plot, 100 random 16-bit sequences are generated, and the average spectrum is displayed.

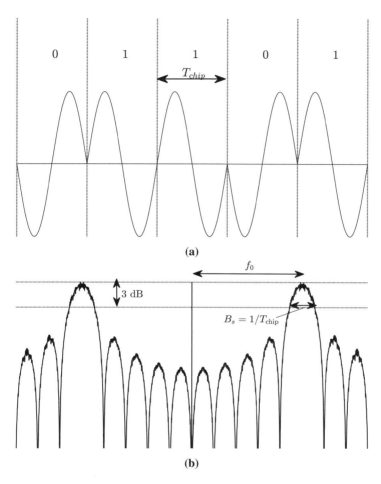

Figure 4.1 Digital modulation with BPSK: (a) 5-bit transmit signal and resultant BPSK baseband waveform, and (b) resultant power spectrum after modulation with a carrier signal at $f_0 = 4 * B_s$; spectrum generated by averaging over 1,000 random transmit codes.

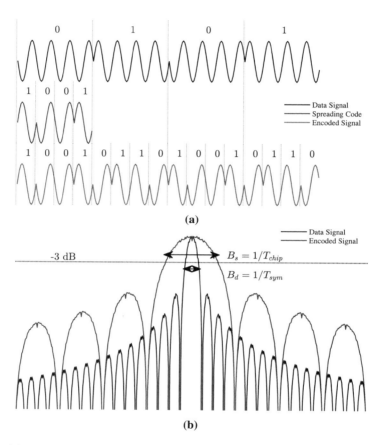

Figure 4.2 Illustration of the spreading process and the relationship between chip rate and bandwidth: (a) Description of a low data rate signal (top), spreading code (middle), and spread spectrum encoded signal (bottom). (b) Illustration of the relative bandwidth of the original (low data rate) signal and encoded (high data rate) signal, with a 4x reduction in chip time, corresponding to a 4x increase in bandwidth.

4.2 FORMULATION

We begin with the same binary hypothesis test as used in the previous chapters.

$$\begin{aligned} \mathcal{H}_0 &: \mathbf{z} = \mathbf{n} & \sim \mathcal{CN}\left(0, \sigma_n^2 \mathbf{I}_M\right) \\ \mathcal{H}_1 &: \mathbf{z} = \mathbf{s} + \mathbf{n} & \sim \mathcal{CN}\left(\mathbf{s}, \sigma_n^2 \mathbf{I}_M\right) \end{aligned} \quad (4.2)$$

4.2.1 DSSS Encoding

DSSS signals are achieved by taking a series of information symbols d_i and multiplying each by some repeated code $c(t)$, which operates at a much higher rate. These codes can either be *long* (the code sequence is longer than a single symbol) or *short* (the code sequence repeats at least once every symbol). In either case, the cooperative receiver uses knowledge of the spreading code to *despread* or *demodulate* the DSSS signal when it is received and recover the narrow bandwidth transmission. The demodulation process provides a boost to the SNR, since the desired signal responds to the demodulation filter, while noise and interfering signals do not; this is referred to as the *processing gain* of DSSS and is given by the equation

$$G_p = \frac{R_s}{R_d} \quad (4.3)$$

where R_s is the chip rate of the spreading sequence (in bits/sec), and R_d is the datarate of the underlying transmission (in bits/sec). Equivalently, if the signal has the same modulation before and after spreading, then the gain can be equivalently expressed as the ratio of bandwidths

$$G_p = \frac{B_s}{B_d} \quad (4.4)$$

See Figure 4.2 for an illustration of the bandwidth before and after spreading is applied.

Operationally, short codes are much easier for an adversary to decode [3], particularly if the code duration T_c is less than the chip rate T_{symobl} (meaning that each symbol contains a copy of the entire coding sequence). Thus, long codes are more secure, but they also increase complexity and delay, since synchronization of the receiver requires correlating the received signal with the known code, to determine the appropriate delay.

DSSS signals, along with issues of detection, geolocation, and jamming, are discussed at a high level in [3, 4]. For more detailed discussion of the mathematics involved, including relative performance of different jamming techniques, consult [5].

4.2.2 Spread Spectrum Radar Signals

LPD signals are frequently employed for radar, although they are often referred to as LPI in that context. The most common class of LPI radar signals are PSK-modulated, in which the phase of the underlying carrier signal is adjusted every T_{chip} seconds.

PSK radar signals are written

$$s(t) = Ae^{j(2\pi f_c t + \phi(t))} \qquad (4.5)$$

for some time series of phase offsets $\phi(t)$. The phase code can be binary (0 and π) or polyphase. Barker and Costas codes are popular choices, due to their sidelobe structure.[2] Increasingly, waveform design is being applied to select an optimal choice of codes for some constraint or desired outcome. See, for example, a recent review of waveform diversity for radar by Blunt and Mokole [6].

The characteristics of radar signals vary with mission, but most are designed to have a *thumb-tack* like ambiguity function, with low sidelobes in the vicinity of the primary target response in both range and Doppler.

4.3 SOLUTION

Since the transmitter is assumed to be noncooperative, the detector has no knowledge of the chip rate, or any other parameters of the signal. This section considers three types of detectors, those that are ignorant of signal structure, those that exploit a property of spread spectrum signals called *cyclostationarity*, and those that leverage multiple receivers to detect coherent signals.

4.3.1 Energy Detectors

The first, and most obvious, solution is to solve the binary hypothesis test directly, with no additional assumptions about the structure of $s(t)$. This results in the same

---

2   See [1], Chapter 5, for a detailed discussion of PSK radar waveforms, including performance and common codes employed.

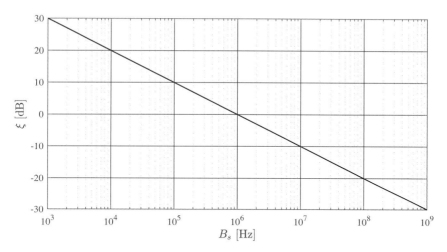

Figure 4.3 Impact of spreading bandwidth on a notional received signal that achieved $\xi = 0$ dB at $B_s = 1$ MHz. For every 10x increase in bandwidth, there is a 10-dB reduction in ξ, caused by the extra noise incident on the receiver.

solution derived in Chapter 3, an energy detector

$$T(\mathbf{z}) = \frac{2}{\sigma_n^2} \mathbf{z}^H \mathbf{z} \qquad (4.6)$$

The equations for P_{FA}, η, and P_{D} are all unchanged. The only difference in this case is that the signal is spread over a much wider bandwidth, which reduces its PSD and corresponding detection performance.

Figure 4.3 shows the SNR for a range of bandwidths, given a reference signal that achieves $\xi = 0$ dB when it occupies $B = 1$ MHz of spectrum. For each 10x increase in bandwidth, there is a corresponding 10x increase in noise power, which causes a 10-dB reduction in ξ. Since the detector does not have knowledge of the spreading sequence used to generate the signal, it can't take advantage of the processing gain G_p to overcome that additional noise.

4.3.2 Cyclostationary Detectors

Stationarity refers to changes in the autocorrelation of a signal over time, which can reflect a periodic component of a signal, such as a tone. Cyclostationarity refers to a

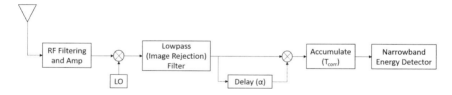

Figure 4.4 Drawing of the DMR.

similar phenomenon within the autocorrelation, namely that the autocorrelation of a signal has some periodic behavior. Put simply, signals that transmit a unique chip every T_{chip} seconds will exhibit periodic behavior in their autocorrelation function over time. To measure this behavior, we define first the *cyclic autocorrelation function* (CAF)

$$R_\alpha(\tau) = \frac{1}{T_0} \int_{-T_0/2}^{T_0/2} R(t; t+\tau) e^{j2\pi\alpha t} dt \qquad (4.7)$$

where $R(t; t+\tau)$ is the autocorrelation function for a signal at time t and with delay τ, and α is the cycle frequency. The Fourier transform of the CAF is called the *spectral correlation function* (SCF)

$$S_\alpha(f) = \int_{-\infty}^{\infty} R_\alpha(\tau) e^{j2\pi f \tau} d\tau \qquad (4.8)$$

This section briefly introduces three types of cyclostationary detectors, but their implementation details are not included. For more discussion of cyclostationarity, including resolution bounds and methods for estimating the CAF and SCF from real signals, readers are referred to [1, 7, 8]. For MATLAB® functions to estimate the SCF, readers are referred to [9].

4.3.2.1 Delay and Multiply Receiver

One option to exploit the periodic structure of spread spectrum signals is the *delay-and-multiply receiver* (DMR). A delay of α is chosen, and the received signal is delayed by that amount, and then multiplied by an undelayed copy. This is shown in Figure 4.4. If there is a cyclic feature corresponding to the delay chosen, then this results in a spectral tone offset from the carrier, which can be detected with a narrowband energy detector [8].

The output of the DMR is written

$$z = \int_0^T y(t)y(t-\alpha)dt \qquad (4.9)$$

Unfortunately, this requires knowledge of the signal structure, such as the chip rate and the modulation employed, as the location of cyclic features will differ based on the modulation. To operate in a noncooperative manner, the DMR must be swept across candidate delays.

Furthermore, the multiplication in a DMR limits operation to the positive SNR regime. If SNR < 0 dB, then the multiplication will increase the noise power more than it increases signal power. This is troublesome, since spread spectrum signals are frequently received with low SNR.

4.3.2.2 Eigendecomposition-Based Receivers

In [8], a pair of DSSS receivers are presented that rely on the eigendecomposition of a sample covariance matrix that is generated from nonoverlapping segments of the received signal. These receivers first estimate several parameters of the DSSS signal, and then make a determination as to its presence from those estimates.

The sample covariance is written

$$R = \frac{1}{K-1} \sum_{k=0}^{K-1} \mathbf{y}_k \mathbf{y}_k^H \qquad (4.10)$$

These algorithms assume that the period of the spreading code is estimated, and the sample windows $\mathbf{y}_k$ have the same length as the spreading code.

In essence, reliance on the eigendecomposition of a sample covariance matrix exploits the fact that noise, being stationary, will result in a sample covariance matrix that is roughly an identity matrix. The eigenvalues of such a matrix are roughly equal. Any cyclostationary signals that are present have a periodic nature, and will lie within one or two eigenvectors (provided that the sample covariance matrix is of the appropriate size). Thus, the size of the first few eigenvectors can indicate the presence of a cyclostationary signal, and the corresponding eigenvectors can be used to estimate the spreading code $c(t)$ that was applied.

The two specific techniques, referred to as *eigenanalysis* and *spectral norm maximization,* differ only in how they process the eigenvectors. Simulations and analysis in [8] show that these methods can detect DSSS signals far below the noise level.

The chief limitation of these techniques is that they only apply to short codes, for which the spreading code repeats at least once for each transmitted symbol. In practice, long codes are much more difficult to descramble [3]. For implementation details, interested readers are referred to Chapter 17 of [8].

4.3.2.3 ML-Based Detector

The cyclostationary detectors discussed so far were developed via heuristics and intuition, not through any formal method of optimization. An optimal detector is discussed in [7], but it is not feasible to implement for noncooperative transmitters, since it relies on perfect knowledge of various signal parameters, such as chip rate. To achieve an operationally realistic detector, an approximation was derived that reduces to integration of the SCF estimated from the received signal (**y**) over a small region around the *cell under test*:

$$T_\alpha(t, f) = \frac{1}{\Delta f} \int_{f-\Delta f/2}^{f+\Delta f/2} S_\mathbf{y}^\alpha(t, v) dv \qquad (4.11)$$

This detector applies to a single cycle frequency α, spectral frequency f, and sample time t (the *cell under test*). To detect noncooperative signals, this test must be repeated for each possible cyclic frequency α and spectral frequency f at each time t that the detector is evaluated.

4.3.3 Cross-Correlation Detectors

Section 4.3.2 describes a class of complex receivers painstakingly designed to exploit structure in the target signal, in an attempt to improve performance over the baseline energy detector. Unfortunately, this results in detectors that are often best for a specific subset of target signals (i.e., radar signals or DSSS transmissions with a given modulation) and require estimation of signal parameters before a detection decision can be made.

Another approach to exploit this structure is to implement a second receiver and employ cross-correlation. Cross-correlation has been studied for LPI radar detection [10–12], and is commonly employed as a means of time-delay estimation for TDOA geolocation [13–15]. The noise terms are assumed to be uncorrelated, so any correlation between the two receivers can be attributed to a signal that was incident on the receivers.

The principal advantage of such a detector is that it can exploit processing gain of the target signals without the need to estimate parameters such as the chip rate. Given an input SNR

$$\xi_i = \frac{S}{N} \quad (4.12)$$

where S and N are the power spectral densities of the two input signals (averaged over the receiver bandwidth), the output SNR after cross-correlation processing is [10, 11]

$$\xi_o = \begin{cases} \frac{T_{\text{corr}} B_n \xi_i^2}{1+2\xi_i} & T_{\text{corr}} \leq T_p \\ \frac{T_p B_n \xi_i^2}{\frac{T_{\text{corr}}}{T_p}+2\xi_i} & T_{\text{corr}} > T_p \end{cases} \quad (4.13)$$

where T_p is the pulse duration and T_{corr} is the correlator's duration (how long the signals are multiplied and the result integrated). B_s is the signal bandwidth, which for simplicity we assume to be matched to the receiver bandwidth (B_n).[3] Figure 4.5 plots this SNR increase for a signal with $T_p = 1\mu s$ pulse width and varying bandwidth. The pulse compression gain of each signal is given by the product of bandwidth and pulse duration [referred to as the *time-bandwidth product* (TBWP)]. The first signal has a TBWP of 1, and thus there is no benefit to correlation processing. The final signal has a TBWP of 1,000, which corresponds to a 30-dB increase in ξ with coherent processing; the dashed lines in Figure 4.5 represent this ideal processing gain that can be achieved with knowledge of the transmit signal. Note that, at high SNR, the cross-correlator approaches, but does not fully achieve, this ideal processing gain. At low SNR, there is more significant roll-off, caused by the increase in noise power from multiplying two noisy signals. If the TBWP is large enough, however, the processing gain can overcome this loss.

Consider a pair of received signals $\mathbf{y}_1$ and $\mathbf{y}_2$. We can write the binary hypothesis test

$$\begin{aligned} \mathcal{H}_0 &: \mathbf{y}_1 = \mathbf{n}_1; \quad \mathbf{y}_2 = \mathbf{n}_2 \\ \mathcal{H}_1 &: \mathbf{y}_1 = \mathbf{s}+\mathbf{n}_1; \quad \mathbf{y}_2 = A\mathbf{s}+\mathbf{n}_2 \end{aligned} \quad (4.14)$$

where A represents the unknown attenuation and phase shift between the two receivers. The cross-correlation receiver is given by

$$R = \mathbf{y}_1^H \mathbf{y}_2 \quad (4.15)$$

Note that this only tests the *zero-lag* term of the cross-correlation. This simplification can be made if the two receivers are very close together, as is often the case for

---

[3] If the receiver noise bandwidth B_n is less than the signal bandwidth B_s, then a loss term should be applied, to reflect the fraction of signal energy that falls outside the receiver's noise bandwidth.

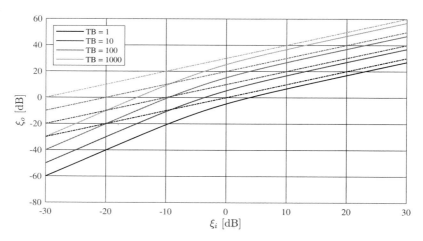

Figure 4.5 Relationship between input and output SNR for a cross-correlation detector. Solid lines represent the cross-correlator SNR output; dashed lines represent an ideal matched filter.

correlators used for signal detection. Those used for time-delay estimation require calculation of multiple lag terms of the cross-correlation. Since both $\mathbf{y}_1$ and $\mathbf{y}_2$ are complex Gaussian random vectors, and their variances are equal, the distribution of their inner product can be found in [16], but it is not given in closed form, nor is it a standard distribution. This is a challenge not only to predicting performance, but also to setting the appropriate threshold to guarantee P_{FA}.

Instead, we rely on an approximation that yields a much more convenient and analyzable solution, with the understanding that there will be a performance penalty and a loss of analytic prediction accuracy. To construct this approximation, we expand the terms of the inner product and consider them individually. There are four terms: a signal cross-correlation, a noise cross-correlation, and cross-terms of the signal with the two noise vectors

$$R = A\mathbf{s}^H\mathbf{s} + \mathbf{s}^H\mathbf{n}_2 + A\mathbf{n}_1^H\mathbf{s} + \mathbf{n}_1^H\mathbf{n}_2 \qquad (4.16)$$

The first term is the (deterministic) signal of interest, while the remaining three are noise terms. The two cross-terms are complex Gaussian random variables with zero mean, and variance given $\sigma_{sn} = \sigma_n^2 ST_p$.[4] The final noise term is more complex;

---

4 This is a straightforward derivation, noting that $\mathbf{s}$ is deterministic and $\mathbf{s}^H\mathbf{s} = ST_p$, and $\mathbf{n}_1 \sim \mathcal{CN}\left(0, \sigma_n^2\right)$, where $\sigma_n^2 = N/B_n$.

it is the inner product of a pair of zero-mean complex Gaussian random variables. If the sample vectors are of sufficient length, however, it is suitable to apply the *central limit theorem* [17] and approximate it with a complex Gaussian distribution, by computing the expectation (0) and variance ($T_p N^2 / B_n$). Rolling up all of these terms, we approximate the distribution of R with

$$R \sim \mathcal{CN}\left(AST_p, \frac{T_p N}{B_n}(N + 2S)\right) \quad (4.17)$$

This leads to the SNR

$$\xi_o = \frac{|A|^2 S^2 T_p^2}{\frac{T_p N}{B_n}(N + 2S)} \quad (4.18)$$

$$= \frac{|A|^2 T_p B_n S^2}{N^2 + 2SN} \quad (4.19)$$

$$= \frac{|A|^2 T_p B_n \xi_i^2}{1 + 2\xi_i} \quad (4.20)$$

This matches the result from [11], under the assumption that $|A|^2 = 1$, and that the received bandwidth B_n is matched to the signal bandwidth B_s.

Both the mean and the variance of R depend on the received signal strength S. This means that both the mean and the variance of R will change between the null and alternative hypothesis, and that the optimal detector is quadratic (see §4.4 of [17]). In the case of spread spectrum signal detection, the signal strength (S) is often much weaker than the noise variance (N). Thus, the variance is approximately constant ($\sigma_r^2 \approx T_p N^2 / B_n$), and the use of a simple threshold test is typically suitable.[5]

To construct a test statistic, we note that the expectation of R depends on the unknown phase shift between the two received vectors. Thus, we must consider a test statistic that is invariant to A. We can accomplish this by taking the magnitude squared of R, divided by the noise under $\mathcal{H}_0$.

$$z(\mathbf{y}_1, \mathbf{y}_2) = \frac{2}{\sigma_0^2} |\mathbf{y}_1^H \mathbf{y}_2|^2 \quad (4.21)$$

---

[5] A truly optimal detector would utilize the likelihood ratio $l(\mathbf{y}_1, \mathbf{y}_2) = f(\mathbf{y}_1, \mathbf{y}_2 | \mathcal{H}_1) / f(\mathbf{y}_1, \mathbf{y}_2 | \mathcal{H}_0)$, but that equation does not simplify easily, and the linear threshold test above lends itself to a more straightforward analysis of the probabilities of false alarm and detection.

where

$$\sigma_0^2 = \frac{N^2 T_{corr}}{B_n} \tag{4.22}$$

This follows a chi-squared distribution with 2 degrees of freedom. Under $\mathcal{H}_1$, the variance of R is written

$$\sigma_1^2 = \sigma_0^2 (1 + 2\xi_i) \tag{4.23}$$

and the probability of false alarm can be computed with

$$P_{\text{FA}} = 1 - F_{\chi^2}(\eta; 2) \tag{4.24}$$

$$\eta = F_{\chi^2}^{-1}(1 - P_{\text{FA}}; 2) \tag{4.25}$$

Next, we define a dummy variable

$$\widehat{z}(\mathbf{y}_1, \mathbf{y}_2) = \frac{\sigma_0^2}{\sigma_1^2} z(\mathbf{y}_1, \mathbf{y}_2) = \frac{z(\mathbf{y}_1, \mathbf{y}_2)}{1 + 2\xi_i} \tag{4.26}$$

which is distributed under $\mathcal{H}_1$ as a noncentral chi-squared random variable with 2 degrees of freedom and noncentrality parameter

$$\lambda = \xi_o \tag{4.27}$$

To compute the probability of detection, we define an equivalent interval for $\widehat{z}(\mathbf{y}_1, \mathbf{y}_2)$

$$Pr\{z(\mathbf{y}_1, \mathbf{y}_2) \geq \eta\} = Pr\left\{\widehat{z}(\mathbf{y}_1, \mathbf{y}_2) \geq \frac{\eta}{1 + 2\xi_i}\right\} \tag{4.28}$$

From here, we can see that the equation to compute the probability of detection is based on the noncentral chi-squared CDF, but with a shifted threshold

$$P_{\text{D}} = 1 - F_{\chi^2}\left(\frac{\eta}{1 + 2\xi_i}; 2, \xi_o\right) \tag{4.29}$$

4.4 PERFORMANCE ANALYSIS

To examine the relative performance of an energy detector and a cross-correlation detector, we consider detection of a commercial wideband cell phone signal, modeled after 3G CDMA waveforms, from an aircraft, as well as detection of an airborne LPI-style radar pulse from a ground receiver on a mast.

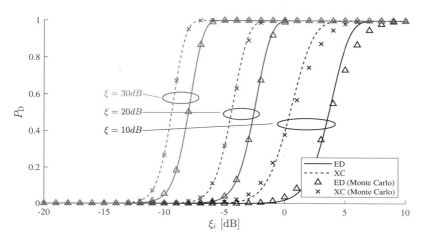

Figure 4.6 Performance comparison of an energy detector (solid lines, triangular marks) and cross-correlation detector (dashed lines, x marks), for various time-bandwidth products.

First, however, we compare the detection performance of the energy detector as outlined in (4.6) with the cross-correlation receiver as defined in (4.21). The result of analytic predictions from the above sections are compared to the Monte Carlo simulations in Figure 4.6.[6] The curves are run for three candidate signals, with time-bandwidth products ranging from 10 dB to 30 dB, across a range of input SNR values (ξ_i).

Note that the performance of the energy detector and cross-correlation are closely matched. This is because these calculations assume that the noise is perfectly known, in which case the energy detector can be well calibrated to detect any increase in overall energy, particularly when M is large. The cross-correlation detector still benefits from the fact that it is comparing correlated signals with uncorrelated noise, while the energy detector has only a single noisy sample of the signal.[7]

---

6   The Monte Carlo simulations are carried out for 10,000 random signal and noise vectors, scaled to the appropriate relative power levels, and run through detectors in the provided MATLAB® package. The code to run this test is found in the examples folder.
7   When considering an estimate of the noise power, the variance terms σ_0 and σ_1 should be replaced with their estimates. The resultant test statistics follow the *student-t* distribution, and in some cases the F distribution [17].

Table 4.1
Parameters for Example 4.1

(a) Transmitter

Parameter	Tower	Handset
P_t	20 W	200 mW
G_t	-3 dBi[9]	0 dBi
f_0	850 MHz	1,900 MHz
h	100m	2m
B_s	5 MHz	
T_p	20 ms	

(b) Receiver

Parameter	Receiver
G_r	0 dBi
L_r	3 dB
F_n	5 dB
B_n	50 MHz
h_r	1,000m
$T_d = T_{corr}$	100 μs

4.4.1 Detection of a 3G CDMA Cell Signal

Consider the parameters in Table 4.1a for both a handset user as well as the base station of a 3G CDMA cell phone network.[8] Parameters for a receiver mounted on a low-flying aircraft are given in Table 4.1b.

Example 4.1 Maximum Detection Range of a 3G CDMA Signal

What is the maximum range at which an energy detector and a cross-correlation detector can detect the tower and the handset? For simplicity, assume that they are the only two users transmitting. Assume that the maximum acceptable P_{FA} is 10^{-6} and that the desired P_D is 0.8.

We begin with a consideration of the detectors. For the energy detector, we compute η and P_D using (3.25) and (3.26), respectively. In the accompanying set of scripts, this is called with

```
> xi_min_ed = detector.squareLawMinSNR(pfa,pd,M);
```

---

8 For details, see [18, 19].
9 This represents a distant sidelobe (-20 dB below peak gain) of the moderately directional 17-dBi peak gain for a nominal cell phone tower transmitter. More likely than not, the receiver will be in the same beam as at least one of the cell phone towers transmitters; since 360° coverage is typically provided.

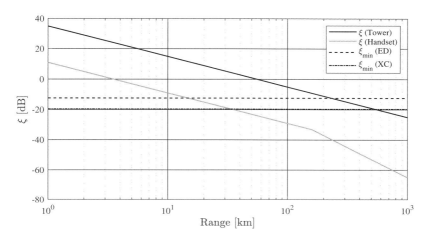

Figure 4.7 Received SNR (ξ) for the cell tower and a handheld user in Example 4.1, as a function of range, compared with the required SNR for both the energy detector and cross-correlator.

where M is the number of samples in the detector ($T_{corr} * B_n$). This results in a minimum SNR of -12.43 dB.

For the cross-correlation detector, we can apply the SNR gain equation from (4.20) to determine that an input SNR of $\xi_i = -19.66$ dB is sufficient to achieve the required -12.43 dB after cross-correlation processing, an improvement of 7.2 dB. This can alternatively be computed with the script

```
> xi_xc = detector.xcorrMinSNR(PFA,PD,Tcorr,Tp,Bn,Bs);
```

Next, we assemble the reference SNR (ξ_0) according to (3.39), which results in 125.98 dB for the tower and 108.98 dB for the user. Then we compute the loss as a function of range and find when the achieved SNR meets the threshold for detection, which is plotted in Figure 4.7. Alternatively, we can use the provided MATLAB® scripts to directly solve for this intersection, with the code

```
> R_ed = detector.squareLawMaxRange(pfa,pd,M,f0,ht,hr,xi0,b@x...
                                    includeAtmLoss,atmStruct);
> R_xc = detector.xcorrMaxRange(pfa,pd,Tcorr,Tp,Bn,Bs,f0,ht,hr,xi0,...
                                    includeAtmLoss,atmStruct);
```

Table 4.2
LPI Radar Parameters for Example 4.2

(a) Transmitter

Parameter	Value
P_t	500W
G_t (mainlobe)	30 dBi
G_t (near sidelobes)	10 dBi
G_t (far sidelobes)	0 dBi
f_0	16 GHz
L_t	3 dB
T_p	20 μs
B_s	200 MHz
h_t	20,000 ft (6,096m)

(b) Detector for an LPI radar pulse

Parameter	Receiver
G_r	0 dBi
L_r	3 dB
F_n	4 dB
B_n	500 MHz
h_r	20m
$T_d = T_{corr}$	100 μs

The energy detector can detect the tower as far as 233 km away, but the handheld user must be no more than 14.8 km away. The cross-correlation detector extends these ranges to 537 km and 34 km, respectively.

4.4.2 Detection of a Wideband Radar Pulse

Consider detection of an LPI radar pulse, whose parameters are given in Table 4.2a. Parameters for a receiver are given in Table 4.2b.

Example 4.2 Maximum Detection Range of an LPI Radar Pulse

What is the maximum detection range for an energy detector and a cross-correlation detector on the provided radar and detector when the radar is (a) pointed at the detector ($G_t = 30$ dBi), (b) pointed near the detector ($G_t = 10$ dBi), or (c) pointed far away from the detector ($G_t = 0$ dBi)? Assume that the maximum acceptable P_{FA} is 10^{-6}, and that the desired P_D is 0.8. We follow a similar approach with the radar pulse as was done with the communications pulse. First the detection thresholds for the energy detector and cross-correlation detector are computed

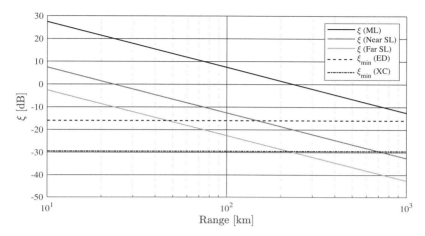

Figure 4.8 Plot of the received SNR as a function of range for the example radar pulse and detector in Example 4.2, compared with the SNR thresholds for the energy detector and cross-correlation detector.

```
> xi_min_ed = detector.squareLawMinSNR(PFA,PD,M);
> xi_xc = detector.xcorrMinSNR(PFA,PD,Tcorr,Tp,Bn,Bs);
```

The result is $\xi = -15.98$ dB for the energy detector and $\xi = -29.49$ dB for the cross correlation detector, a gain of 13.5 dB. Next, we compute the received SNR as a function of range, plotted in Figure 4.8 for mainlobe detection, near-in sidelobe detection, and distant sidelobe detection.

From Figure 4.8, we can see that the energy detector can observe the radar pulse at a range of 1,482 km (mainlobe), 149 km (near sidelobes), and 46.9 km (far sidelobes). The cross-correlation detector can extend those maximum ranges to 7,025 km, 702.5 km, and 222 km, respectively.

4.5 LIMITATIONS

A principal limitation of the detection of spread spectrum signals is that the SNR can be made prohibitively low by the transmitter, simply by spreading the energy transmitted over a tremendously large bandwidth. While energy detectors and cross-correlation receivers with a long enough observation window can recover some of that spreading gain, it still requires sampling a very large noise bandwidth. While some techniques exist that greatly exploit the structure in spread spectrum signals (i.e., cyclostationary detectors), they are often reliant on knowledge or estimation of signal parameters for successful operation.

Another limitation, particularly for wideband signals, is that cross-correlators have to consider a large number of possible delays, which can involve a significant computational effort.

4.6 PROBLEM SET

4.1 Link-16 (a military datalink developed in the 1980s) utilizes DSSS encoding with a chip rate of 32 chips in each 6.4-μs pulse. Based on this, what is the instantaneous bandwidth of a Link-16 transmission?

4.2 Each 6.4-μs Link-16 pulse contains 5 information bits but is encoded with 32 chips. What is the DSSS processing gain?

4.3 The public GPS waveform is referred to as *coarse/acquisition* (C/A). It is transmitted at a rate of 1.023×10^6 chips per second. What is the bandwidth of the C/A code? The underlying data message is transmitted at a rate of 50 bps. What is the DSSS processing gain?

4.4 The median received signal strength for C/A code transmissions is -154 dBW at the Earth's surface (assuming an omnidirectional receive antenna). If the receiver has a 4-MHz bandwidth, 1 dB of receive loss, and a 2-dB noise figure, what is the SNR for a receiver that has no knowledge of the signal code? What is the SNR if the receiver knows the DSSS sequence?

4.5 Consider the noncoherent detection of the encrypted GPS Y-code signal, which has a bandwidth of 10.23 MHz and initial SNR of $\xi = -20$ dB. What output SNR would a cross-correlation detector achieve with a sampling interval of $T_p = 1$ s?

4.6 For the cross-correlator in the previous problem, what is the P_D if the detector threshold is set for $P_{FA} = 10^{-3}$?

References

[1] P. E. Pace, *Detecting and classifying low probability of intercept radar.* Norwood, MA: Artech House, 2009.

[2] M. Skolnik, *Radar Handbook, 3rd Edition.* New York, NY: McGraw-Hill Education, 2008.

[3] D. Adamy, *EW 103: Tactical battlefield communications electronic warfare.* Norwood, MA: Artech House, 2008.

[4] D. Adamy, *EW 104: EW Against a New Generation of Threats.* Norwood, MA: Artech House, 2015.

[5] R. Poisel, *Modern Communications Jamming: Principles and Techniques, 2nd Edition.* Norwood, MA: Artech House, 2011.

[6] S. D. Blunt and E. L. Mokole, "Overview of radar waveform diversity," *IEEE Aerospace and Electronic Systems Magazine*, vol. 31, no. 11, pp. 2–42, November 2016.

[7] A. M. Gillman, "Non Co-Operative Detection of LPI/LPD Signals Via Cyclic Spectral Analysis." Master's thesis, Air Force Institute of Technology, 1999.

[8] R. A. Poisel, *Electronic warfare receivers and receiving systems.* Norwood, MA: Artech House, 2015.

[9] E. L. d. Costa, "Detection and identification of cyclostationary signals." Master's thesis, Naval Postgraduate School, 1996.

[10] A. W. Houghton and C. D. Reeve, "Spread spectrum signal detection using a cross correlation receiver," in *1995 Sixth International Conference on Radio Receivers and Associated Systems*, Sep 1995, pp. 42–46.

[11] A. W. Houghton and C. D. Reeve, "Detection of spread-spectrum signals using the time-domain filtered cross spectral density," *IEE Proceedings - Radar, Sonar and Navigation*, vol. 142, no. 6, pp. 286–292, Dec 1995.

[12] R. Ardoino and A. Megna, "LPI radar detection: SNR performances for a dual channel cross-correlation based ESM receiver," in *2009 European Radar Conference (EuRAD)*, Sept 2009.

[13] I. Jameson, "Time delay estimation," Defence Science and Technology Organization, Electronic Warfare and Radar Division, Tech. Rep., 2006.

[14] W. R. Hahn, "Optimum passive signal processing for array delay vector estimation," Naval Ordnance Laboratory, Tech. Rep., 1972.

[15] W. Hahn and S. Tretter, "Optimum processing for delay-vector estimation in passive signal arrays," *IEEE Transactions on Information Theory*, vol. 19, no. 5, pp. 608–614, 1973.

[16] M. K. Simon, *Probability distributions involving Gaussian random variables: A handbook for engineers and scientists*.   New York, NY: Springer Science & Business Media, 2007.

[17] L. L. Scharf, *Statistical signal processing*.   Reading, MA: Addison-Wesley, 1991.

[18] D. N. Knisely, S. Kumar, S. Laha, and S. Nanda, "Evolution of wireless data services: Is-95 to cdma 2000," *IEEE Communications Magazine*, vol. 36, no. 10, pp. 140–149, 1998.

[19] V. K. Garg, *IS-95 CDMA and CDMA2000: Cellular/PCS systems implementation*.   New York, NY: Pearson Education, 1999.

Chapter 5

Scanning Receivers

Chapters 3 and 4 discussed, respectively, the detection of an unknown target signal that is either continuous and narrowband, or spread across the electromagnetic spectrum. In each of these cases, it was assumed that the target signal's carrier frequency was centered within the bandwidth of the receiver.

In reality, there is a very broad swath of the electromagnetic spectrum that must be observed. Radar signals can span from the single tens of megahertz, with high frequency (HF) over the horizon radar systems, up to the millimeter-wave regime (30–300 GHz) that is increasingly occupied by automotive radars in the commercial world. Within that span, there are military radar examples that can be found to occupy nearly every band.

Similarly, on the communications side, long-haul voice links can be found across HF (3–30 MHz), VHF (30–300 MHz), and UHF (300 MHz–1 GHz), and digital communications (via datalink) are spread across the low gigahertz regime; Link-16 occupies L band (1–2 GHz), and many UAV datalinks span S band (2–4 GHz) and C band (4–8 GHz). Satellite up and down links occupy many higher frequencies including Ku band (12–18 GHz), Ka band (27–40 GHz) and V band (40–75 GHz).[1]

Even a relatively focused receiver, such as one dedicated to isolating Link-16 transmissions, must cover a fairly wide swath of frequencies, as Link-16 can hop rapidly among 51 different 5-MHz-wide channels that span from 900 MHz to 1,206 MHz, for a hopping bandwidth of 306 MHz, far larger than the 5-MHz channel

---

1 Within this text, we follow the IEEE band designations, which give a letter to each frequency between 1 and 110 GHz. The International Telecommunications Union and NATO both have competing frequency band designations.

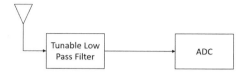

Figure 5.1 Notional RF digital receiver.

bandwidth in use at any one point in time.[2] For radar threats, it is not uncommon with modern digital systems to have a radar that, for example, can operate over a several gigahertz frequency band (such as between 8 and 12 GHz) but occupy as little as 1 MHz for a low-resolution detection mode, or as much as 1 GHz for a high-resolution imaging or classification mode. In either case, only a small portion of the observed spectrum may be in use at any given time, but it is unknown to the receiver where to focus their efforts.

Historically, EW systems have employed a wide variety of specialized receivers. Increasingly, however, the pace of commercial technology has enabled digital systems with ever-increasing fidelity and bandwidth. For simplicity, this text focuses on digital receivers. Interested readers are referred to Chapter 4 of [1] (particularly Table 4.1) for a general treatment of the most important analog receivers, including their relative strengths and weaknesses, and to Chapters 8 and 9 of [2] for a technical derivation of the two most prominent EW receiver architectures: superheterodyne and compressive. Some specific receiver architectures (and expected nominal performance) are also discussed in the Naval Warfare Center's *EW Handbook* [3]. Poisel goes into detail on the search for, detection, and jamming of communications signals [4]. For radar, different waveforms are often employed; these are discussed at length in [5].

5.1 DIGITAL RECEIVERS

Digital receivers are preferable to analog receivers for a host of reasons, including lower cost and ever-increasing performance (as a result of advances in the commercial electronics and wireless telecommunications sectors), as well as the ability to process data in multiple algorithms or channels without the need to split power, thereby preserving SNR. A simplified block diagram is shown in Figure 5.1. The

---

2   Many of the 51 channels may be simultaneously in use, but any given transmitter will occupy only one at a time.

heart of a digital receiver is the analog-to-digital converter (ADC), which converts input signal voltage to a digital representation.

If the ADC's sampling rate exceeds the Nyquist frequency for the highest possible target frequency, then a direct digital sampling is possible, as is shown in Figure 5.1. This is a straightforward design. The low-pass filter serves to prevent higher frequency signals from entering the receiver, where they would be aliased down into the band of interest.[3] Without the need for a *local oscillator* (LO), the preselection filter can be replaced with a low-pass filter to prevent aliasing during digitization. This simplifies the cost and complexity of the receiver but has limited application at higher frequencies.

5.1.1 In-Phase (I) and Quadrature (Q) Digitization

Signals can be digitized in one of two forms; either using real samples alone or via complex sampling to obtain the I and Q samples. When sampling digitally, the Nyquist criterion is relaxed, as the complex samples provide the necessary information to disambiguate signals when sampled at the highest frequency of the received signal, as opposed to twice the highest frequency (the standard Nyquist sampling criterion).[4]

Figure 5.2 shows the architectures for a complex ADC. The input signal is mixed with both the sine and cosine of a local oscillator at the center frequency of the bandwidth to be sampled. Each of the two branches are then passed to an ADC, and the outputs collected as the real (I) and imaginary (Q) values of the complex valued sample. We assume for simplicity that complex sampling is utilized.

5.2 IF RECEIVERS

If the highest frequency of interest exceeds the sampling rate of the ADC, then direct sampling is not possible, and an alternative approach is required, such as the *intermediate frequency* (IF) receiver shown in Figure 5.3.[5] IF receivers are based

---

3 See [6] for background reading on digital sampling, including the Nyquist sampling criteria and aliasing.
4 Despite this relaxation, many EW systems (particularly DRFM jammers) choose to operate at greater than twice the Nyquist rate. This is done largely to compensation for the effects of quantization on signal reconstruction [1].
5 This diagram is rather simplistic; real systems often include ancillary components such as amplifiers and attenuators and may include multiple stages of filtering and down-conversion. Interested readers are referred to [2] for a more rigorous discussion of receiver components and architectures.

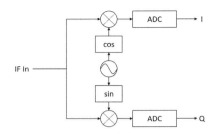

Figure 5.2 I and Q digitization.

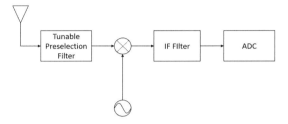

Figure 5.3 Notional IF digital receiver.

on the principle that two signals, when mixed, will result in combined signals at the sum and difference of the two inputs signals, sometimes called their *beat frequencies*. This is a result of the cosine identity

$$\cos(a)\cos(b) = \frac{1}{2}\left[\cos\left(a-b\right) + \cos\left(a+b\right)\right] \quad (5.1)$$

Thus, an incoming signal at very high frequency, can be mixed with a local oscillator, resulting in a signal that has a very low center frequency. This basic principle allows us to take signals transmitted on any carrier frequency for which we can build antennas and filters and mix it down to a much lower frequency to allow for fine-grained channel selection and detection. This process was first patented in 1901 and has been in wide use ever since [2]. It also forms the heart of a *superheterodyne* receiver, one of the most widely employed analog receiver architectures.

The preselection filter is designed to reject ambiguous signals, called *images* before the mixing stage, while the IF filter prevents aliasing in the ADC. If the signal was properly filtered, then it can now be run through an ADC that is run at a much lower rate given by the bandwidth of the IF filter, rather than the maximum frequency of the input signal. This is one of the most important benefits of IF receivers. The IF filter electronics and detector need not run at the RF frequency of interest (which may be in the tens or hundreds of gigahertz), but can be designed to accept signals between DC and the instantaneous bandwidth of the IF filter.

Example 5.1   Frequency Hopping Signal Detection

Consider a frequency hopping communications signal with a hop rate of $f_{\text{hop}} = 100$ hops/s ($T_{\text{hop}} = 10$ ms), instantaneous bandwidth of $B_t = 5$ MHz, and hopping bandwidth of $B_{\text{hop}} = 200$ MHz.

If we assume that the receiver's bandwidth is equal to the signal bandwidth ($B_r = 5$ MHz), and that we want to scan the entire channel at least once per hop to detect the signal, how long can we dwell on each channel, and what probability of detection is achievable (assuming a maximum acceptable P_{FA} of 10^{-6})?

We begin by computing how many different frequencies must be sampled to scan the entire hopping bandwidth:

$$N_{\text{chan}} = \left\lfloor \frac{B_{\text{hop}}}{B_r} \right\rfloor = 40 \tag{5.2}$$

and compute the available time for each frequency sample, or dwell, within the scan

$$T_{\text{dwell}} = \frac{T_{\text{hop}}}{N_{\text{chan}}} = 250 \ \mu s \tag{5.3}$$

At the sampling rate of 5 MHz, each frequency sample will contain $M = 1,250$ samples. This detection curve is shown in Figure 5.4 (solid line). To achieve a detection probability of, say 90%, requires SNR of $\xi \geq -7.4$ dB. If, instead, the signal hops at a rate of T_{hop}=100 hops/sec (dashed line) or 1,000 hops/sec (dotted line), as some modern military datalinks do, similar detection performance would require SNR levels of $\xi \geq -1.8$ dB or 4.7 dB, respectively.

All of these curves assume that the only job of the detector is to find the hopping signal, not to perform any analysis. If countermeasures are required, such as jamming the signal as it hops, then the scan period must be a fraction of the hopping period, rather than equivalent to the full length.

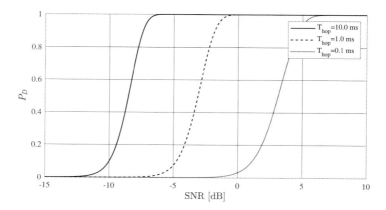

Figure 5.4 Performance of a scanning IF receiver as described in Example 5.1.

5.3 FREQUENCY RESOLUTION

A second constraint on scanning receivers is the desired frequency resolution, denoted δ_f. Once the signal is digitized, it can be processed via the discrete Fourier transform, which will result in frequency bins whose bandwidth is $\delta_\omega = 1/M$. Mapping this to real frequency units, we have [6]

$$\delta_f = \frac{1}{T_{\text{dwell}}} \qquad (5.4)$$

This, along with the receiver's instantaneous bandwidth, presents a fundamental limitation. For a given channel size, such as the 5-MHz channel size of Link-16 transmissions, a minimum dwell time is enforced (200 ns, in the case of a Link-16 transmission). If, for example, our receiver is limited to $B_r = 60$ MHz, then we'll need six scans to cover the full Link-16 band ($B_{\text{hop}} = 306$ MHz). At 200 ns per dwell, it will take 1.2 μs to cover the Link-16 band. This is sufficient to detect every frequency hop, since Link-16 has a hop rate of $T_{\text{hop}} = 76.9$ μs. This does not take into account the signal energy and whether the dwell time is sufficient to actually detect the presence of the signal on a given channel, only that it is possible to scan the frequency channels involved and determine which channel is in use on any given hop.

An example is shown in Figure 5.5, for a case where the dwell time is too large, and the receiver is not able to scan the hopping bandwidth quickly enough

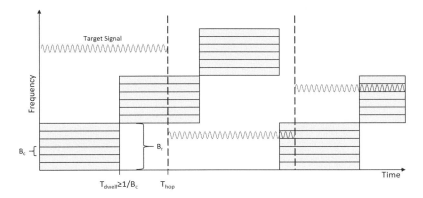

Figure 5.5 Example scanning digital channelized receiver, showing the interaction of dwell time and hopping period.

to detect every pulse. Three separate scans are required in this example in order to cover the band of interest. This means that the scan period is $T_{\text{scan}} = 3T_{\text{dwell}}$. This is much larger than the hopping period T_{hop}; thus the receiver is not guaranteed to detect every hop. Additionally, of those hops that do overlap with the scan strategy, many pulses are only partially intercepted, which will necessarily reduce their probability of detection.

Example 5.2 FMCW Radar

Consider a frequency-modulated continuous wave (FMCW) radar that dwells on a given frequency for $T_{\text{hop}} = 1$ ms, occupies $B_s = 10$ MHz of bandwidth, and may hop over any frequency between 8 and 12 GHz ($B_{\text{hop}} = 4$ GHz).

What receiver bandwidth is required to adequately find this signal on every hop if we desire a frequency resolution equal to the transmit bandwidth ($\delta_f = B_s$)? What if we desire $\delta_f = 100$ kHz?

First, we compute how long we must dwell on each frequency in order to obtain a resolution at least as small as the radar's channel size $B_s = 10$ MHz. This results in $T_{\text{dwell}} \geq 100$ ns. From this, the number of scans that can be tolerated is

$$N_{\text{dwell}} = \left\lfloor \frac{T_{\text{hop}}}{T_{\text{dwell}}} \right\rfloor = 10,000 \qquad (5.5)$$

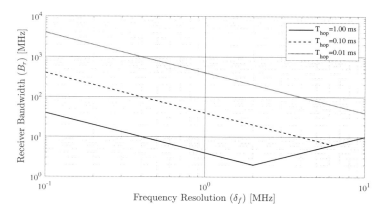

Figure 5.6 Required receiver bandwidth (B_r) as a function of the desired frequency resolution (δ_f) for Example 5.2.

We can compute the allowable receiver bandwidth as the ratio of band of interest (such as the target signal's hopping bandwidth) divided by the number of scans that can fit within the hopping period. The receiver bandwidth should also be at least as large as the desired frequency resolution.

$$B_r = \max\left(\delta_f, \frac{B_{\text{scan}}}{N_{\text{dwell}}}\right) \quad (5.6)$$

In this case, the result is $B_r = \delta_f = 10$ MHz. Figure 5.6 plots the required receiver bandwidth as a function of desired channel size for various hopping rates. Note that, as the desired frequency resolution increases, the required receiver bandwidth decreases (to a point). This is because the larger frequency resolution dictates a shorter dwell time and allows more scans to occur within the required hopping period. This inflection occurs when $\delta_f = B_{\text{scan}}/N_{\text{dwell}}$; beyond that point δ_f begins the driving force and (eventually) B_r is independent of the hopping period.

At the specified hopping rate, $T_{\text{hop}} = 1$ ms, we can confirm the calculation already discussed, $B_r \geq 10$ MHz. To achieve a frequency resolution of $\delta_f = 100$ kHz, the receiver bandwidth must now be $B_r \geq 40$ MHz. Notice that, as the hopping period T_{hop} decreases, the bandwidth must increase. This is because the receiver has fewer collection opportunities (N_{dwell}), and so each dwell must collect a larger portion of the hopping band.

Recall that these calculations ignore the detectability constraint on dwell time (the minimum dwell time required to achieve the desired P_D for a given P_{FA}).

Example 5.3 Pulsed Frequency Agile Radar

Consider a pulsed radar with a PRF of 2 kHz (PRI = 500 μs) and 20% duty cycle ($t_p = 125$ μs), the same spectral characteristics (transmit and hopping bandwidth) as Example 5.2, and the ability to select a new frequency on each pulse.

What receiver bandwidth is necessary to detect each pulse with a desired frequency resolution of $\delta_f = 10$ MHz? What if $\delta_f = 100$ kHz?

The approach to this problem is the same as before, except that we must scan the entire band of interest during the pulse duration. Since each pulse will select a new frequency, the receiver must be able to scan the entire band of interest in t_p seconds. In this case, $t_p = 125$ μs. Just as in the previous example, the dwell period is $T_{\text{dwell}} = 100$ ns, and the resultant number of dwells allowed is $N_{\text{dwell}} = 1,250$.

From here, the computation for required receiver bandwidth is the same as in the previous example, and the results are plotted in Figure 5.7 for three different pulse durations, and a number of desired frequency resolutions. From this figure, we see that the required bandwidth for $\delta_f = 10$ MHz (and $t_p = 125\mu s$) is $B_r \geq 10$ MHz, while for $\delta_f = 100$ kHz, this increases to $B_r \geq 200$ MHz. If the pulse duration is further shortened, the dwell duration lengthens and the required receiver bandwidth increases correspondingly.

5.4 PROBLEM SET

5.1 Consider a frequency-hopping communications signal with a hop rate of $f_{\text{hop}} = 1,000$ hops/s, an instantaneous bandwidth of $B_t = 10$ MHz, and a hopping bandwidth of $B_{\text{hop}} = 500$ MHz. Assume that the receiver bandwidth is 10 times the instantaneous bandwidth $B_r = 100$ MHz, and that we scan the entire channel at least once per hop to detect the signal. How long can we dwell on each channel, and what probability of detection is achievable (assuming a maximum acceptable P_{FA} of 10^{-6} and per-sample input SNR $\xi = -15$ dB).

5.2 Consider a push-to-talk radio occupying a 10-kHz channel in the band 150–160 MHz. If each transmission is expected to last $T = 5$ seconds, what is the maximum acceptable dwell time for a scanning receiver with $B_r = 100$ kHz?

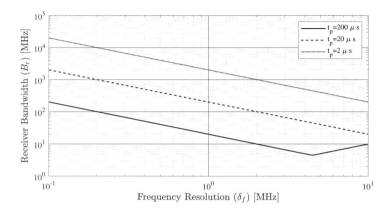

Figure 5.7 Required receiver bandwidth (B_r) as a function of the desired frequency resolution (δ_f) for Example 5.3.

5.3 For the push-to-talk radio in Problem 5.2, how many samples can be collected at each frequency channel (assume Nyquist sampling rate)? What is the achievable P_D if the P_{FA} is set to 10^{-3} and the per sample SNR is $\xi = -13$ dB?

5.4 Consider a frequency hopping radar with a pulse duration of $\tau_p = 50$ μs, bandwidth $B_t = 10$ MHz, and hopping bandwidth $B_{\text{hop}} = 2$ GHz. What is the minimum dwell time to ensure that the proper frequency is identified with $\delta_f \leq 10$ MHz? From this dwell period, what is the minimum receiver bandwidth B_r to ensure that each pulse is detected?

5.5 Repeat the previous problem with a desired frequency resolution of $\delta_f \leq 100$ kHz.

5.6 Consider an FMCW radar that dwells on a given frequency for $T_{\text{hop}} = 10$ ms, occupies $B_s = 250$ MHz of bandwidth, and may hop over any frequency between 16 and 22 GHz. What receiver bandwidth is required to adequately find this signal on every hop, if we desire frequency resolution equal to the transmit bandwidth ($\delta_f = B_s$)? What if we desire $\delta_f = 100$ kHz?

5.7 Consider a pulsed radar with a PRF of 10 kHz and 15% duty cycle. The signal bandwidth is $B_s = 250$ MHz, and hopping bandwidth is 6 GHz, and each pulse selects a new frequency. What receiver bandwidth is necessary to detect each pulse with a desired frequency resolution of $\delta_f = 10$ MHz? What if $\delta_f = 100$ kHz?

References

[1] D. Adamy, *EW 103: Tactical battlefield communications electronic warfare*. Norwood, MA: Artech House, 2008.

[2] R. A. Poisel, *Electronic warfare receivers and receiving systems*. Norwood, MA: Artech House, 2015.

[3] Avionics Department, *Electronic Warfare and Radar Systems: Engineering Handbook*, 4th ed. China Lake, CA: Naval Air Warfare Center Weapons Division, 2013.

[4] R. Poisel, *Modern Communications Jamming: Principles and Techniques, 2nd Edition*. Norwood, MA: Artech House, 2011.

[5] P. E. Pace, *Detecting and classifying low probability of intercept radar*. Norwood, MA: Artech House, 2009.

[6] A. V. Oppenheim, A. S. Willsky, and S. H. Nawab, *Signals and Systems*, 2nd ed. New Jersey: Prentice Hall, 1997.

Part II
Angle of Arrival Estimation

Chapter 6

Estimation Theory

While detection of emitters is required for general situational awareness, it is also the first step to any response action. The second piece of information that is often needed is from what direction the adversary is approaching. To do this, we must estimate some parameters of the detected signal, often either the direction or AOA, or the physical location from which the signal emanates. This chapter discusses the general basis of estimation theory. The principles of estimation theory will be utilized later when we discuss AOA techniques in Chapters 7 and 8, as well as the precision geolocation techniques in Chapters 10–13.

6.1 BACKGROUND

Estimation theory is the mathematical underpinning of how measurements are processed to obtain an estimate of some unknown parameter. It is built on the same statistical foundations as detection theory (discussed in Chapter 2) and has a number of approaches, each optimal in a different sense.

We consider first some measurement vector $\mathbf{x}$, which is described by the probability distribution function $f(\mathbf{x}|\vartheta)$, where ϑ is a set of unknown parameters that modify the distribution. An estimator is some function that computes an estimate $\widehat{\vartheta}$ from the observed data $\mathbf{x}$ and implicit knowledge of the distribution $f(\mathbf{x}|\vartheta)$.

First, this chapter introduces the concept of *maximum likelihood* (ML) estimation, which computes the parameter ϑ that is most likely to have resulted in the observation $\mathbf{x}$. We then briefly review some other prominent estimation strategies and discuss performance metrics for estimation, particularly as concerns the estimation of the AOA of a signal.

6.2 MAXIMUM LIKELIHOOD ESTIMATION

Maximum likelihood estimation is based around the simple concept of finding the parameter estimate $\widehat{\vartheta}$ that maximizes the likelihood of the observed data $\mathbf{x}$. Mathematically, this is written

$$\widehat{\vartheta} = \arg\left[\max_{\vartheta} f(\mathbf{x}|\vartheta)\right] \qquad (6.1)$$

This topic is covered in detail in Chapter 6 of [1]; interested readers are directed there for a more thorough discussion. This section is intended to be a brief review. If the likelihood function is continuous, then the maximization process can be solved by setting the derivative (with respect to each element of the parameter vector ϑ) equal to zero. This gradient is referred to as the *score function* $s(\vartheta, \mathbf{x})$:

$$s(\vartheta, \mathbf{x}) = \frac{\partial}{\partial \vartheta} \ln\left[f(\mathbf{x}|\vartheta)\right] \qquad (6.2)$$

Example 6.1 Sample Mean and Sample Variance

Consider a sample vector $\mathbf{x}$ with M elements, where each element is independent and identically distributed as a Gaussian random variable with mean μ and variance σ^2. Both the mean and variance are unknown, thus the parameter vector is

$$\vartheta = \begin{bmatrix} \mu \\ \sigma^2 \end{bmatrix} \qquad (6.3)$$

Find the score function $s(\vartheta, \mathbf{x})$ and solve to obtain the ML estimate of ϑ.

We begin with the probability distribution function of a complex Gaussian random vector that has independent and identically distributed elements with complex mean μ and variance σ^2:

$$f(\mathbf{x}|\mu, \sigma^2) = (\pi\sigma^2)^{-M} e^{-\frac{1}{\sigma^2}\sum_{i=1}^{M}|x_i - \mu|^2} \qquad (6.4)$$

and the log likelihood function

$$\ell\left(\mathbf{x}|\mu,\sigma^2\right) = -M\ln\left(\pi\sigma^2\right) - \frac{1}{\sigma^2}\sum_{i=1}^{M}|x_i - \mu|^2 \tag{6.5}$$

The score function is the derivative of (6.5) with respect to the two parameters μ and σ^2.

$$\mathbf{s}\left(\mathbf{x}, \boldsymbol{\vartheta}\right) = \begin{bmatrix} \frac{\partial \ell(\mathbf{x}|\mu,\sigma^2)}{\partial \mu} \\ \frac{\partial \ell(\mathbf{x}|\mu,\sigma^2)}{\partial \sigma^2} \end{bmatrix} \tag{6.6}$$

With some algebraic manipulation, we can solve for the two partial derivatives:[1]

$$\mathbf{s}\left(\mathbf{x}, \boldsymbol{\vartheta}\right) = \begin{bmatrix} \frac{1}{\sigma^2}\sum_{i=1}^{M}(x_i - \mu)^* \\ \frac{-M}{(\sigma^2)^2}\left[\sigma^2 - \frac{1}{M}\sum_{i=1}^{M}|x_i - \mu|^2\right] \end{bmatrix} \tag{6.7}$$

To solve for $\widehat{\mu}$ and $\widehat{\sigma^2}$, we replace μ and σ^2 in (6.7) with their estimates, and then set the vector equal to $\mathbf{0}$ and solve.

$$\begin{bmatrix} \frac{1}{\sigma^2}\sum_{i=1}^{M}(x_i - \widehat{\mu})^* \\ \frac{-M}{(\widehat{\sigma}^2)^2}\left[\widehat{\sigma}^2 - \frac{1}{M}\sum_{i=1}^{M}|x_i - \widehat{\mu}|^2\right] \end{bmatrix} = \begin{bmatrix} 0 \\ 0 \end{bmatrix} \tag{6.8}$$

Solving each equation in turn leads to the ML estimates:

$$\widehat{\mu} = \frac{1}{M}\sum_{i=1}^{M}x_i \tag{6.9}$$

$$\widehat{\sigma}^2 = \frac{1}{M}\sum_{i=1}^{M}|x_i - \widehat{\mu}|^2 \tag{6.10}$$

---

[1] For a complex variable z, the derivative $\frac{\partial}{\partial z}|z|^2 = z^*$. This is called a *Wirtinger derivative*. For more information on complex calculus, see [2, 3]. Appendix 2 of [3] contains a table of common derivatives.

6.3 OTHER ESTIMATORS

6.3.1 Minimum Variance Unbiased Estimators

Minimum Variance Unbiased Estimators are similar to maximum likelihood estimators, but are intentionally formulated to have no bias in their estimates. This is very useful for performance metrics, as many of the statistical performance bounds (such as the CRLB, discussed in Section 6.4.2) do not apply to biased estimators.

For example, take the expectation of the ML estimator for the variance of a Gaussian random variable, derived in Section 6.2.

$$E\left[\widehat{\sigma}^2\right] = E\left[\frac{1}{M}\sum_{i=1}^{M}|x_i - \widehat{\mu}|^2\right] \tag{6.11}$$

$$= \frac{1}{M}\sum_{i=1}^{M} E\left[|x_i|^2 - 2\Re\{x_i\widehat{\mu}^*\} + |\widehat{\mu}|^2\right] \tag{6.12}$$

$$= \frac{1}{M}\sum_{i=1}^{M} |\mu|^2 + \sigma^2 - 2\left(|\mu|^2 + \frac{\sigma^2}{M}\right) + \left(|\mu|^2 + \frac{\sigma^2}{M}\right) \tag{6.13}$$

$$= \frac{M-1}{M}\sigma^2 \tag{6.14}$$

From this, we can see that while (6.10) is the maximum likelihood estimate of σ^2, it is not unbiased. In this case, since the estimate $\widehat{\sigma}$ is a scaled version of the true estimate, we can easily construct an unbiased estimator

$$\widehat{\sigma}^2_{UB} = \frac{1}{M-1}\sum_{i=1}^{M}|x_i - \widehat{\mu}|^2 \tag{6.15}$$

Proving that an unbiased estimator is the minimum variance unbiased estimator is not a straightforward process. The first step is to find an estimator and compare its variance to the CRLB (Section 6.4.2). If the variance of an unbiased estimator achieves this bound, then it is said to be *efficient* and is guaranteed to be an MVUB (although, not necessarily the *only* MVUB). The problem is more challenging if the unbiased estimator is not efficient. There is no elegant solution to proving that an unbiased estimator is minimum variance if it is no efficient, short of testing all known unbiased estimators.

6.3.2 Bayes Estimators

While maximum likelihood estimators work on the principle of finding the estimate that is most likely to have resulted in the observed data, Bayes estimators work by minimizing the risk of an incorrect answer. This is predicated upon two new functions, the loss function $L\left(\widehat{\vartheta}, \vartheta\right)$ and the prior (the *a priori* distribution of possible parameters ϑ), $f(\vartheta)$. This section briefly introduces Bayes estimation theory. Interested readers are directed to Chapter 7 of [1].[2]

The loss function can take any form desired, but one popular (and useful) example is the euclidean distance:

$$L\left(\vartheta, \widehat{\vartheta}\right) = \left(\vartheta - \widehat{\vartheta}\right)^H \left(\vartheta - \widehat{\vartheta}\right) \tag{6.16}$$

where $(\cdot)^H$ is the Hermitian (or conjugate transpose).

The prior distribution $f(\vartheta)$ can take on any form, but is most useful if it is continuous. A uniform prior distribution represents the situation where little *a priori* information is known.

From the prior and the loss function, the Bayes risk is computed

$$R\left(\widehat{\vartheta}\right) = \int \int L\left(\vartheta, \widehat{\vartheta}\right) f(\mathbf{x}, \vartheta) \, d\mathbf{x} \, d\vartheta \tag{6.17}$$

where $f(\mathbf{x}, \vartheta)$ is the joint distribution over $\mathbf{x}$ and ϑ, defined as the product of the likelihood function $f(\mathbf{x}|\vartheta)$ and the prior distribution $f(\vartheta)$.

$$f(\mathbf{x}, \vartheta) = f(\mathbf{x}|\vartheta) f(\vartheta) \tag{6.18}$$

With this expression of Bayes risk, we can now formally define the Bayes estimator

$$\widehat{\vartheta}_B = \arg\left[\min_{\widehat{\vartheta}} R\left(\widehat{\vartheta}\right)\right] \tag{6.19}$$

Bayes estimators are most useful when there is some a priori information, such as an initial estimate of ϑ from some other source.

---

[2] Recall that the estimator $\widehat{\vartheta}(\mathbf{x})$ is a function of the data vector $\mathbf{x}$. For simplicity, we often omit this fact and write $\widehat{\vartheta}$ in isolation.

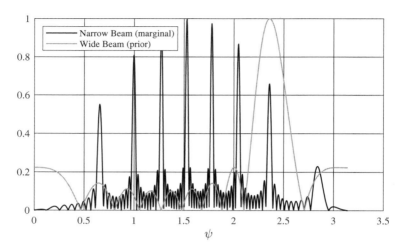

Figure 6.1 Example conditional and prior distributions as a function of ψ used in Examples 6.2 and 6.3.

Example 6.2   Bayes Estimation Example

Consider the case of a phased array being used for direction of arrival estimation. The details are discussed in Chapter 8, but the array pattern for an array whose broadside is pointed at $\psi = \pi/2$, and whose individual elements have a $\cos(\psi)^{1.2}$ response, from a signal whose direction of arrival is ψ_0 is given

$$f(x|\psi) = |\cos(\psi - \pi/2)|^{1.2} \left| \frac{\sin\left(\pi N \frac{d}{\lambda}\left(\cos(\psi) - \cos(\psi_0)\right)\right)}{\sin\left(\pi \frac{d}{\lambda}\left(\cos(\psi) - \cos(\psi_0)\right)\right)} \right| \quad (6.20)$$

Now, we take one such pattern, with $d/\lambda = 4$, $N = 10$ and peaks and grating lobes as seen in Figure 6.1 for the narrow beam (marginal) case. What is the Bayes estimate if we assume no prior information? In other words, assume that the prior distribution is

$$f(\psi) = \begin{cases} 1/\pi & 0 \leq \psi \leq \pi \\ 0 & \text{otherwise} \end{cases} \quad (6.21)$$

It is clear from an analysis of Figure 6.1 that the peak occurs in the vicinity of $\psi = \pi/2$, and that is (by definition) the maximum likelihood estimate $\widehat{\theta}$. Since the Bayesian prior $f(\psi)$ provides no information, the joint distribution $f(\mathbf{x}, \theta)$ is equivalent to the marginal $f(\mathbf{x}|\theta)$, and we must conclude that the ML estimate is also the Bayesian estimate. $\widehat{\psi} = \pi/2$.

Example 6.3 Bayes Estimation with Informative Prior

Next, consider that this phased array was cued by a smaller array, one that also has $N = 10$ elements, but has a narrower interelement spacing of $d/\lambda = .5$. The pattern for this beam is shown in Figure 6.1 as the wide beam (prior) case. If this information is considered as our prior, then what is the Bayesian estimate of the angle of arrival?

In this case, the Bayesian prior is the wide peak centered on $\psi \approx 2.35$. The joint distribution is taken as the product of the two curves in Figure 6.1. Without going into the math, it is clear that the peak of the joint distribution will fall near $\widehat{\psi} \approx 2.35$. In this manner, the Bayesian estimator allows us to consider any prior information that is available. It need not be from a cueing sensor, as discussed here, and could be from common sense (such as the fact that there is a mountain pass at a given heading, so any ground units are likely coming through there), or from off-board intelligence, such as an overhead ISR cue that suggests the adversary is to the northeast of the current position. If the information can be expressed in the form of a prior distribution, it can be incorporated into the Bayesian estimate.

6.3.3 Least Square Estimators

In least squares estimation, the goal is to minimize the sum of the squared error term. Let $g(\vartheta)$ be a model that describes the relationship between ϑ and $\mathbf{x}$, such that

$$\mathbf{x} \approx \mathbf{g}(\vartheta) \tag{6.22}$$

If we plug our estimate into the function, then the error can be measured

$$\epsilon\left(\widehat{\vartheta}\right) = \left[\mathbf{x} - \mathbf{g}\left(\widehat{\vartheta}\right)\right]^H \left[\mathbf{x} - \mathbf{g}\left(\widehat{\vartheta}\right)\right] \tag{6.23}$$

The least squares estimator takes the value ϑ from the optimization function

$$\widehat{\vartheta}_{LS} = \arg\left[\min_{\widehat{\vartheta}} \epsilon\left(\widehat{\vartheta}\right)\right] \tag{6.24}$$

If the function $\mathbf{g}(\vartheta)$ is linear, then we can rewrite it $\mathbf{G}\vartheta$, and the solution to the optimization problem is:

$$\widehat{\vartheta}_{LS} = \left(\mathbf{G}^H \mathbf{G}\right)^{-1} \mathbf{G}^H \mathbf{x} \qquad (6.25)$$

6.3.4 Convex Estimators

If the optimization problem, such as (6.24), can be shown to be convex (that is, if the cost function to be optimized is convex, and the constraints are all convex), then a number of fast numerical solvers can be utilized to arrive at a solution. The most popular are variations of linear or nonlinear least squares solutions and gradient descent algorithms. In each of these approaches, an initial estimate of the parameter ϑ is made, and the local region around that estimate is examined to generate a direction in which the exact solution is believed to be. The algorithm then iteratively "steps" in the direction that is believed to be toward the exact solution, and the process is repeated. Many algorithms can converge quickly in a short number of steps, but it is not uncommon to require hundreds or thousands of algorithm iterations to arrive at a solution.

Figure 6.2 shows a notional two-dimensional convex optimization problem. The rings form a contour map, and the △ is the true optimal point. The dashed line and + markers indicate the results of a number of iterative steps in the optimization algorithm (beginning with an initial, or seed, estimate); each one computes the direction of steepest descent in the region near the current estimate and then computes an updated solution some distance along that direction. After several iterations, the estimate is very close to the true solution. This approach, and subtle variations, are useful for rapid solutions to convex problems, for which an analytical solution is not feasible.

Interested readers are referred to [4] for a detailed discussion of convex problems and numerical methods to solve them efficiently. We will deal with these methods briefly in Chapter 11.

6.3.5 Tracking Estimators

Tracking estimators, such as the Kalman filter, use repeated observations to track a parameter over time. At each point in time the tracker constructs not only an estimate of the current value, but a prediction of what that parameter will be some short time in the future. Then, when the next observation is made, the prediction

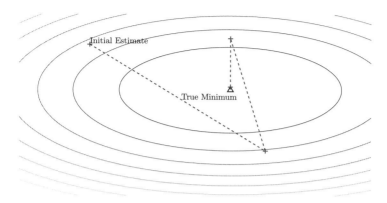

Figure 6.2 A notional convex optimization problem. Concentric rings form a contour plot, while the dashed line marks progress from the initial estimate (x) through each successive iteration (+) toward the true optimal solution (△).

is updated, and another prediction is generated. These filters often require a system model (such as a kinematic model for tracking a target's position), and measurement model (translating the parameters true values to noise measurements and estimates). In this text, we do not consider the need to continually track a target's estimated position, but that is a very important problem, which cannot be overlooked when implementing an EW geolocation system, since targets on a battlefield are rarely stationary.

Figure 6.3 shows a few time steps of a notional tracking problem. At each time step, the 1σ uncertainty interval of the prediction (based on all prior data) is shown in the shaded region. The measurement collected at that time instant is shown as an x, while the smoothed position estimate (which considers both the measurement and the prior distribution) is marked with an o. Over time, the prediction interval gets smaller, as the measurements appear within the predicted interval, increasing confidence in the position estimate over time. However, when an aberration occurs, this is reflected with a growth in the uncertainty on the following step.

A brief derivation of the Kalman filter is found in Section 7.8 of [1]. For nonlinear systems, exceptions such as the extended Kalman filter, unscented Kalman filter, and particle filter have been well studied [5, 6].

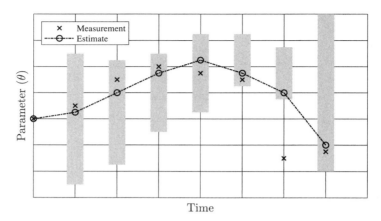

Figure 6.3 An example tracking problem with *a priori* uncertainty intervals, measurements, and updated tracker estimates.

6.4 PERFORMANCE MEASURES

For any of the estimators described so far, the estimate $\widehat{\vartheta}$ is properly viewed as a random vector, since it is the result of a deterministic function (the estimator) applied to a random vector (the observation $\mathbf{x}$). Performance can be described by analyzing the distribution of the error term

$$\boldsymbol{\epsilon} = \widehat{\vartheta} - \vartheta \tag{6.26}$$

Most commonly, we analyze the first and second moments of the distribution of $\boldsymbol{\epsilon}$, referred to as the bias and error covariance, respectively

$$\mathbf{b} = E\{\boldsymbol{\epsilon}\} = E\left\{\widehat{\vartheta} - \vartheta\right\} \tag{6.27}$$

$$\mathbf{C} = E\left\{(\boldsymbol{\epsilon} - \mathbf{b})(\boldsymbol{\epsilon} - \mathbf{b})^H\right\} \tag{6.28}$$

For an *unbiased* estimator, $\mathbf{b} = \mathbf{0}$, and we can rewrite the error covariance

$$\mathbf{C} = E\left\{\left(\widehat{\vartheta} - \vartheta\right)\left(\widehat{\vartheta} - \vartheta\right)^H\right\} \tag{6.29}$$

Most performance measures rely on either the bias or the error covariance.

6.4.1 Root Mean Squared Error (RMSE)

A common error calculation in evaluating estimators is the RMSE. It is commonly defined for a set of sample errors, as in a Monte Carlo trial or experimental test results

$$\text{RMSE} = \sqrt{\frac{1}{N} \sum_{i=1}^{N} \left\| \widehat{\vartheta}_i - \vartheta \right\|^2} \quad (6.30)$$

where $\widehat{\vartheta}_i$ is the ith estimate of ϑ. The error term $\left\| \widehat{\vartheta}_i - \vartheta \right\|^2$ is the square of the ℓ-2 norm, or Euclidean distance, between the estimate and the true position.[3]

6.4.2 CRLB

If a system exists, it can be tested or simulated to obtain an estimate of the error covariance matrix (**C**) and bias (**b**), and any of the measures above can be computed. The downside is that is typically a slow and costly process and does not easily extend to situations that were not tested. Statistical bounds, however, can typically be solved analytically, allowing for insight and intuition into the behavior of the underlying algorithm and providing performance guarantees across a greater range of scenarios.

One popular, and very useful, performance measure is the CRLB. This is a lower bound on the error covariance **C** for an unbiased estimator:

$$\mathbf{C}\left(\widehat{\vartheta}\right) \geq CRLB \quad (6.31)$$

The reason the CRLB is very useful is that it is often achievable and can thus be used to guarantee that a given unbiased estimator is *efficient* (meaning that there is no alternative estimator with a lower error covariance for a given input SNR). Since there is no unbiased estimator that can violate (6.31), any unbiased estimator that achieves it is optimal in that sense.

To compute the CRLB, we start with the score function, which was previously defined as the gradient of the log likelihood function, with respect to the unknown parameter vector ϑ.

$$s\left(\vartheta, \mathbf{x}\right) = \frac{\partial}{\partial \vartheta} \ell\left(\mathbf{x} | \vartheta\right) \quad (6.32)$$

---

3 The ℓ_2 norm is defined $\|\mathbf{a}\| = \sqrt{\mathbf{a}^H \mathbf{a}}$.

The *Fisher information matrix* (FIM) is given by the expectation of the score matrix, multiplied by its own complex conjugate:

$$\mathbf{F}_\vartheta(\mathbf{x}) = E\left[\mathbf{s}(\mathbf{x}, \vartheta)\mathbf{s}^H(\mathbf{x}, \vartheta)\right] \tag{6.33}$$

The CRLB is then given by the inverse of the FIM:

$$C\left(\widehat{\vartheta}\right) \geq \mathbf{F}_\vartheta^{-1}(\mathbf{x}) \tag{6.34}$$

For a more complete definition, including treatment of nuisance parameters, and alternate formulations of the Fisher information matrix, interested readers are referred to Sections 6.4 and 6.5 of [1] or Section 6.3 of [3].

Example 6.4 CRLB of Gaussian with Unknown Mean

Consider the sample vector $\mathbf{x}$, distributed as $x_i \sim \mathcal{CN}\left(\mu, \sigma_n^2\right)$, where μ is unknown, and σ_n^2 is the known noise variance. Compute the CRLB of μ and compare it to the error of $\widehat{\mu}$ in (6.9).

Since μ is unknown, but σ_n^2 is known, we can take $s(x, \mu)$ from the first row of (6.7):

$$s(\mathbf{x}, \mu) = \frac{1}{\sigma^2} \sum_{i=1}^{M} (x_i - \mu)^* \tag{6.35}$$

The FIM is given as:

$$\begin{aligned} F_\mu(\mathbf{x}) &= E\left[\left(\frac{1}{\sigma_n^2}\sum_{i=1}^{M}(x_i - \mu)^*\right)\left(\frac{1}{\sigma_n^2}\sum_{k=1}^{M}(x_k - \mu)^*\right)^*\right] & (6.36) \\ &= \frac{1}{(\sigma_n^2)^2}\sum_{i=1}^{M}\sum_{k=1}^{M} E\left[x_i^* x_k - x_i^* \mu - x_k \mu^* + |\mu|^2\right] & (6.37) \\ &= \frac{1}{(\sigma_n^2)^2}\sigma_n^2 = \frac{M}{\sigma_n^2} & (6.38) \end{aligned}$$

Thus, for any estimator $\widehat{\mu}$:

$$C(\mu) = E\left[|\widehat{\mu} - \mu|^2\right] \geq \frac{\sigma^2}{M} \tag{6.39}$$

We compare this with the error covariance of the sample mean $\hat{\mu}$ from (6.9):

$$E\left[|\hat{\mu} - \mu|^2\right] = E\left[\left|\frac{1}{M}\sum_{i=1}^{M} x_i - \mu\right|^2\right] \qquad (6.40)$$

$$= \frac{1}{M^2}\sum_{i=1}^{M}\sum_{j=1}^{M} E\left[(x_i - \mu)(x_j - \mu)^*\right] = \frac{\sigma_n^2}{M} \qquad (6.41)$$

Since the error covariance of the sample mean in (6.9) is equivalent to the CRLB (and the estimator is unbiased), then we can say that it is *efficient*.

6.4.2.1 Nuisance Parameters

Consider the case where multiple parameters in the vector ϑ are unknown, but only some of them are necessary to estimate. One example would be a situation where the incoming signal has unknown amplitude and phase, but only the phase is of interest (as is the case with many geolocation techniques). Even though some of the unknown parameters are not needed, the fact that they are not known adds uncertainty to the parameters being estimated. To capture this increased uncertainty, we construct and invert the Fisher information matrix using the full parameter vector, and then inspect the components of the CRLB that are related to the parameters of interest.

For example, let us consider the parameter vector ϑ as two distinct sets of parameters, those that are nuisance (ϑ_n) and those that are of interest (ϑ_i).

$$\vartheta = \begin{bmatrix} \vartheta_n^T & \vartheta_i^T \end{bmatrix}^T \qquad (6.42)$$

The Fisher information matrix can then be expressed in block matrix form

$$\mathbf{F}_\vartheta(\mathbf{x}) = \begin{bmatrix} \mathbf{A} & \mathbf{B} \\ \mathbf{C} & \mathbf{D} \end{bmatrix} \qquad (6.43)$$

where $\mathbf{A}$ and $\mathbf{D}$ are the Fisher information matrix of ϑ_n and ϑ_i, respectively, and $\mathbf{B}$ and $\mathbf{C}$ are cross-terms. Then, we can apply the matrix inversion lemma to compute

$\mathbf{F}_\vartheta(\mathbf{x})^{-1}$ in a block fashion [1].

$$\mathbf{F}_\vartheta^{-1}(\mathbf{x}) = \begin{bmatrix} \mathbf{F}_{\vartheta_n}^{-1}(\mathbf{x}) & \mathbf{F}_{\vartheta_n,\vartheta_i}^{-1}(\mathbf{x}) \\ \mathbf{F}_{\vartheta_i,\vartheta_n}^{-1}(\mathbf{x}) & \mathbf{F}_{\vartheta_i}^{-1}(\mathbf{x}) \end{bmatrix} \quad (6.44)$$

$$\mathbf{F}_{\vartheta_n}^{-1}(\mathbf{x}) = \left(\mathbf{A} - \mathbf{B}\mathbf{D}^{-1}\mathbf{C}\right)^{-1} \quad (6.45)$$

$$\mathbf{F}_{\vartheta_n,\vartheta_i}^{-1}(\mathbf{x}) = -\mathbf{A}^{-1}\mathbf{B}\left(\mathbf{D} - \mathbf{C}\mathbf{A}^{-1}\mathbf{B}\right)^{-1} \quad (6.46)$$

$$\mathbf{F}_{\vartheta_i,\vartheta_n}^{-1}(\mathbf{x}) = -\left(\mathbf{D} - \mathbf{C}\mathbf{A}^{-1}\mathbf{B}\right)^{-1}\mathbf{C}\mathbf{A}^{-1} \quad (6.47)$$

$$\mathbf{F}_{\vartheta_i}^{-1}(\mathbf{x}) = \left(\mathbf{D} - \mathbf{C}\mathbf{A}^{-1}\mathbf{B}\right)^{-1} \quad (6.48)$$

Since we only care about ϑ_i, we can compute only $\mathbf{F}_{\vartheta_i}^{-1}(\mathbf{x})$. Thus, the CRLB is written

$$C_{\vartheta_i} \geq \left(\mathbf{D} - \mathbf{C}\mathbf{A}^{-1}\mathbf{B}\right)^{-1} \quad (6.49)$$

We can see from (6.49) that the CRLB of the desired parameters ϑ_i depends on the full Fisher information matrix of ϑ, including the nuisance parameters.

6.4.2.2 CRLB of Functions

Consider a function of the estimated parameter vector $\mathbf{y} = f(\vartheta)$. While it is possible to compute the CRLB of $\mathbf{y}$ directly, a simplification is provided. See Equations 6.102–6.105 in [1], paraphrased here briefly. We define the vectors $\mathbf{G}$ and $\mathbf{H}$ with the gradients of $\mathbf{y}$ as a function of ϑ, and the inverse

$$[\mathbf{G}]_{i,j} = \frac{\partial \vartheta_j}{\partial y_i} \quad (6.50)$$

$$[\mathbf{H}]_{i,j} = \frac{\partial y_j}{\partial \vartheta_i} \quad (6.51)$$

Given these vectors, the CRLB of $\mathbf{y}$ can be computed from the CRLB of ϑ directly, using either $\mathbf{G}$ or $\mathbf{H}$, whichever is most convenient

$$\mathbf{F}_\mathbf{y}^{-1}(\mathbf{x}) = \mathbf{G}^{-H}\mathbf{F}_\vartheta^{-1}(\mathbf{x})\mathbf{G}^{-1} = \mathbf{H}^H \mathbf{F}_\vartheta^{-1}(\mathbf{x})\mathbf{H} \quad (6.52)$$

6.4.2.3 Special Case: Real Gaussian Distribution

Solutions to many common estimation problems abound, but one particularly important result is the estimation of a vector ϑ of parameters from a data set that is Gaussian distribution with real-valued mean vector **μ** and covariance matrix **C**. From [1], the Fisher information matrix is given as:

$$[\mathbf{F}_\vartheta(\mathbf{x})]_{ij} = \frac{1}{2}\text{Tr}\left[\mathbf{C}^{-1}\frac{\partial \mathbf{C}}{\partial \theta_i}\mathbf{C}^{-1}\frac{\partial \mathbf{C}}{\partial_j}\right] + \left(\frac{\partial \mathbf{\mu}}{\partial \theta_i}\right)^T \mathbf{C}^{-1}\left(\frac{\partial \mathbf{\mu}}{\partial \theta_j}\right) \quad (6.53)$$

where $[\mathbf{A}]_{ij}$ denotes the element of the matrix **A** in the ith row and jth column. If the input data vector **x** is sampled M different times,[4] then the Fisher information matrix is multiplied by a scalar M.

6.4.2.4 Special Case: Real Parameters of Complex Gaussian Distribution

Another important special case is the estimation of real-valued parameters from a vector of data that is distributed as a complex Gaussian random variable with mean vector **μ** and covariance matrix **C**. From [7], the Fisher information matrix is given as:

$$[\mathbf{F}_\vartheta(\mathbf{x})]_{ij} = \text{Tr}\left[\mathbf{C}^{-1}\frac{\partial \mathbf{C}}{\partial \theta_i}\mathbf{C}^{-1}\frac{\partial \mathbf{C}}{\partial_j}\right] + 2\Re\left\{\left(\frac{\partial \mathbf{\mu}}{\partial \theta_i}\right)^H \mathbf{C}^{-1}\left(\frac{\partial \mathbf{\mu}}{\partial \theta_j}\right)\right\} \quad (6.54)$$

6.4.2.5 Finding an Efficient Estimator

Conveniently, there is an approach to derive an efficient estimator, *if it exists*. It can be proved that if $\widehat{\theta}$ is efficient *if and only if* [8]

$$\mathbf{F}_\vartheta(\mathbf{x})\left(\widehat{\vartheta} - \vartheta\right) = \frac{\partial}{\partial \vartheta^*}\ell(\mathbf{x}|\vartheta) \quad (6.55)$$

This presents a straightforward approach to finding an efficient estimator. If $s(\mathbf{x}, \vartheta)$ can be manipulated to take the form $\mathbf{F}_\vartheta(\mathbf{x})\left(\widehat{\vartheta} - \vartheta\right)$ for some estimator $\widehat{\vartheta}$, then that estimator is efficient.

Example 6.5 Deriving an Efficient Estimator

For a complex Gaussian random vector **x**, with elements distributed $x_i \sim \mathcal{N}(\mu, \sigma^2)$, with known noise variance σ^2. Derive an efficient estimator for μ.

---

[4] That is, if the data matrix **X** is a repeated set of measurements: $\mathbf{X} = [\mathbf{x}_1, \mathbf{x}_2, \ldots, \mathbf{x}_M]$ and $\mathbf{x}_i \sim \mathcal{N}(\mathbf{m}, \mathbf{C})$).

Similar to the derivation of the score function, we compute the derivative of the log likelihood with respect to μ^*:

$$\frac{\partial}{\partial \mu^*} \ell(\mathbf{x}|\mu) = \frac{1}{\sigma^2} \sum_{i=1}^{M} (x_i - \mu) \qquad (6.56)$$

Recall from (6.38) that $F_\mu(\mathbf{x}) = M/\sigma^2$. Next, we manipulate (6.56) and attempt to isolate a term equal to $F_\mu(\mathbf{x})$. If the remainder takes the form $(\widehat{\mu} - \mu)$ for some estimator $\widehat{\mu}$, then that estimator is efficient. Thus,

$$\begin{aligned}
\frac{\partial}{\partial \mu^*} \ell(\mathbf{x}|\mu) &= \frac{M}{\sigma^2} \left(\frac{1}{M} \sum_{i=1}^{M} (x_i - \mu) \right) & (6.57) \\
&= F_\mu(\mathbf{x}) \left(\frac{1}{M} \sum_{i=1}^{M} x_i - \frac{1}{M} \sum_{k=1}^{M} \mu \right) & (6.58) \\
&= F_\mu(\mathbf{x}) \left(\widehat{\mu} - \mu \right) & (6.59)
\end{aligned}$$

Thus, $\widehat{\mu}$ is efficient. Of course, we already prove this in Section 6.4.2, but this presents a way to derive such an efficient estimator from the score function and Fisher information matrix.

6.4.2.6 Other Statistical Bounds

The CRLB is not the only statistical bound that is relevant to geolocation performance, but it is the most widely cited; this is largely due to its ease of derivation and computation, as well as its achievability.

Another bound that is relevant to geolocation accuracy is the ZZB [9–15]. The ZZB frames the estimation problem as a binary hypothesis test between the true parameter value and a slightly perturbed one, and then computes the smallest perturbation for which the test can successfully discriminate. The advantage of the ZZB is that it is much tighter at moderate and low SNR levels than the CRLB. (The CRLB is overly optimistic when SNR is low.)

A third bound that is also relevant is the Bhattacharyya bound [12, 16, 17], which is more complex but can provide a lower bound on estimation performance for *biased* estimators, while the CRLB and ZZB both apply only to *unbiased* estimators.

Estimation Theory

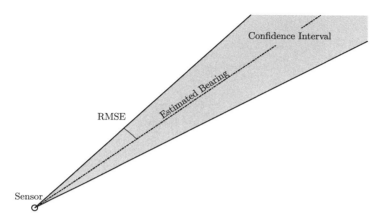

Figure 6.4 Illustration of the estimated bearing and RMSE for an AOA sensor. If it is unbiased, then the shaded region has a 68.26% of containing the source of the detected signal.

6.4.3 Angle Error Variance and Confidence Intervals

In Chapters 7 and 8, we will discuss estimation of the AOA for incoming signals. The principals of RMSE and the CRLB apply directly to this problem. This section briefly discusses how they are applied to angular estimates, and introduces the idea of a confidence interval.

If the errors in the estimated angle of arrival $\widehat{\phi}$ are assumed to be Gaussian-distributed, and if $\widehat{\phi}$ is unbiased, then the RMSE, defined in (6.30) is equivalent to the standard deviation (σ) of $\widehat{\phi}$, and the region defined by $\widehat{\phi} \pm \sigma$ has a 68.26% chance of containing the true estimate σ. An example of this is shown in Figure 6.4.

Using the Gaussian distribution, we can construct arbitrary confidence intervals, or even determine the confidence that an estimate lies within a given bound, using the integral of the standard normal distribution from $-\gamma$ to γ:

$$CI = \frac{1}{\sqrt{2\pi}} \int_{-\gamma}^{\gamma} e^{-\frac{1}{2}z^2} \, dz \qquad (6.60)$$

where γ is the multiplier to apply to the RMSE when setting the confidence interval, and CI is the probability that the interval contains the true estimate (between 0 and 1). This is represented in MATLAB® with:

```
> confInterval = normcdf(gamma) - normcdf(-gamma);
```

Table 6.1
RMSE Scale Factor and Confidence Intervals

Confidence Interval	Scale Factor γ
50%	0.6745
68.26%	1.000
75%	1.1503
90%	1.6449
95%	1.96

The inverse problem is solved:

```
> gamma = norminv(.5 + confInterval/2);
```

This is enumerated for several useful values in Table 6.1. This scaling factor breaks down for large RMSE estimators, since the error is bounded by ± 180 degrees (while a true Gaussian random variable is unbounded).[5]

The CRLB is useful for situations where a specific estimator may not exist or is not known to be optimal. Since the CRLB is a lower bound on error covariance, its square root is a lower bound on the standard deviation (or, equivalently, the RMSE)

$$\text{RMSE} \geq \sqrt{\text{CRLB}} \qquad (6.61)$$

Just as the RMSE can be scaled for various confidence intervals, as shown in (6.60) and Table 6.1, so can the CRLB be scaled to show the bounds on various confidence intervals.

Example 6.6 AOA Accuracy

Assume that there is a sensor with sufficient directionality that the CRLB of AOA estimate is given by

$$\text{CRLB} = 57.3 \frac{\lambda}{D} \cos \phi \qquad (6.62)$$

What is the 50% confidence interval for an angle of arrival measurement of $\widehat{\phi} = 30°$, assuming that $\widehat{\phi}$ is efficient and that $D/\lambda = 10$? What if the error covariance is double the CRLB?

---
[5] If $\sigma > 45°$, then the approximation quickly breaks down. But, for standard deviations below $\sigma = 45°$, the Gaussian approximation is sufficiently accurate.

First, we compute the CRLB at $\widehat{\phi} = 30°$,

$$\text{CRLB} = 57.3 \frac{\cos(30)}{10} = 6.62° \qquad (6.63)$$

From Table 6.1, we see that the multiplier γ for a 50% confidence interval is $\gamma = 0.6745$. Thus, the 50% confidence interval is defined by $\widehat{\phi} \pm 4.46°$ around the true estimate; in other words there is a 50% chance that

$$\phi \in [25.54°, 34.46°] \qquad (6.64)$$

If the error covariance of $\widehat{\phi}$ is twice the CRLB, then we can simply double the confidence interval. In that case, there is a 50% chance that[6]

$$\phi \in [21.1°, 38.9°] \qquad (6.65)$$

6.5 PROBLEM SET

6.1 What is the score function $s(\mathbf{x}, \lambda)$ for a sample vector $\mathbf{x}$ that follows an exponential distribution

$$f(\mathbf{x}|\lambda) = \lambda^{-N} e^{-\lambda \sum_{n=0}^{N-1} x_n}$$

Find the maximum likelihood estimate of the variance $\vartheta = \lambda^2$.

6.2 In Problem 6.1, what is the CRLB on the estimate of ϑ?

6.3 Prove that the ML estimate in Problem 6.1 is not efficient.

6.4 Is the sample variance in (6.15) an efficient estimator? Note that the variable

$$z = \sum_{n=0}^{N-1} \frac{|x_n - \widehat{\mu}|^2}{\sigma_n^2/2} \qquad (6.66)$$

for sample mean $\widehat{\mu}$ and x_n distributed as a complex Gaussian with mean μ and variance σ^2, is distributed as a chi-squared random variable with 2(N-1) degrees of freedom.

---

[6] The $\in$ operator stands for "is contained in the interval."

6.5 Consider a sample vector **x** with N IID samples $x_i \sim \mathcal{CN}(a/b, 1)$. What is the CRLB for estimation error of $|a|^2$, assuming that $|b|^2$ is known.

6.6 Find an efficient estimator for a in the prior problem.

6.7 Consider a sample vector $\overline{\mathbf{x}}$ with N IID elements of $x_i \sim \mathcal{CN}(\mu, \sigma^2)$. Both μ and σ^2 are unknown. Find the CRLB of σ^2; in this case μ is a nuisance parameter. Compare to the CRLB in the case that σ^2 is known.

6.8 Find the 90% confidence interval for a direction-finding estimate of $32°$ that is distributed as a Gaussian random variable with a standard deviation of $2°$.

References

[1] L. L. Scharf, *Statistical signal processing*. Reading, MA: Addison-Wesley, 1991.

[2] R. Hunger, "An introduction to complex differentials and complex differentiability," Munich University of Technology, Inst. for Circuit Theory and Signal Processing, Tech. Rep. TUM-LNS-TR-07-06, 2007.

[3] P. J. Schreier and L. L. Scharf, *Statistical signal processing of complex-valued data: the theory of improper and noncircular signals*. Cambridge, UK: Cambridge University Press, 2010.

[4] S. Boyd and L. Vandenberghe, *Convex optimization*. Cambridge, UK: Cambridge University Press, 2004.

[5] S. J. Julier and J. K. Uhlmann, "Unscented filtering and nonlinear estimation," *Proceedings of the IEEE*, vol. 92, no. 3, pp. 401–422, March 2004.

[6] B. Ristic, S. Arulampalam, and N. Gordon, *Beyond the Kalman filter: Particle filters for tracking applications*. Norwood, MA: Artech House, 2003.

[7] S. M. Kay, *Fundamentals of Statistical Signal Processing, Volume 2: Detection Theory*. Upper Saddle River, NJ: Prentice Hall, 1998.

[8] S. Trampitsch, "Complex-valued data estimation," Master's thesis, Alpen-Adria University of Klagenfurt, 2013.

[9] J. Ziv and M. Zakai, "Some lower bounds on signal parameter estimation," *IEEE Transactions on Information Theory*, vol. 15, no. 3, pp. 386–391, May 1969.

[10] K. L. Bell, Y. Steinberg, Y. Ephraim, and H. L. Van Trees, "Extended ziv-zakai lower bound for vector parameter estimation," *IEEE Transactions on Information Theory*, vol. 43, no. 2, pp. 624–637, Mar 1997.

[11] N. Decarli and D. Dardari, "Ziv-zakai bound for time delay estimation of unknown deterministic signals," in *2014 IEEE International Conference on Acoustics, Speech and Signal Processing (ICASSP)*, May 2014, pp. 4673–4677.

[12] Y. Wang and K. Ho, "TDOA positioning irrespective of source range," *IEEE Transactions on Signal Processing*, vol. 65, no. 6, pp. 1447–1460, 2017.

[13] Y. Wang and K. C. Ho, "Unified near-field and far-field localization for AOA and hybrid AOA-TDOA positionings," *IEEE Transactions on Wireless Communications*, vol. 17, no. 2, pp. 1242–1254, Feb 2018.

[14] D. Khan and K. L. Bell, "Explicit ziv-zakai bound for analysis of doa estimation performance of sparse linear arrays," *Signal Processing*, vol. 93, no. 12, pp. 3449 – 3458, 2013, special Issue on Advances in Sensor Array Processing in Memory of Alex B. Gershman.

[15] D. Dardari, A. Conti, U. Ferner, A. Giorgetti, and M. Z. Win, "Ranging with ultrawide bandwidth signals in multipath environments," *Proceedings of the IEEE*, vol. 97, no. 2, pp. 404–426, Feb 2009.

[16] A. Fend, "On the attainment of cramer-rao and bhattacharyya bounds for the variance of an estimate," *The Annals of Mathematical Statistics*, pp. 381–388, 1959.

[17] J. S. Abel, "A bound on mean-square-estimate error," *IEEE Transactions on Information Theory*, vol. 39, no. 5, pp. 1675–1680, Sep 1993.

Chapter 7

Direction-Finding Systems

AOA information can allow a commander to direct additional surveillance resources or to prepare defenses, potentially thwarting a flank attack. It can also be used to direct airborne interceptors of missiles with seekers that can search for and acquire targets. But, AOA information is most useful for directing electronic attacks.

Direction-finding systems are used to find the AOA of an incoming signal.[1] They have been employed since the early days of radar and wireless communication, to determine an adversary's position. The increasingly common use of digital receivers (discussed in Chapter 5) has led to an improvement in signal quality and performance, but has not greatly affected the architectures or approach of dedicated direction finding systems.

This chapter discusses several purpose-built direction-finding systems. This includes systems based on the amplitude response of a directional antenna or multiple antennas, changes in the Doppler response of a moving antenna, and phase comparison at two antennas. These systems are all purpose-built for direction finding. In Chapter 8, we will introduce methods to perform direction finding from a general purpose antenna array.

This chapter treats many of the direction-finding systems only briefly, but attempts to use consistent formulation for ease of comparison. For another brief review, with operational context, see Dave Adamy's EW 101 series, specifically Chapter 7 of EW 103 [1] and Section 6.6 of EW 104 [2]. Within each section, reference is made to more detailed texts, often with implementation details and

---

1 Direction-finding, angle of arrival, direction of arrival, line of bearing, and position fix are all related terms that are largely synonymous. In this text, we use direction finding (DF) to refer to the systems, architectures, and processes, while AOA is used to refer to the actual estimates.

performance trade-offs. Another good reference is [3], which deals with some of these receivers in Section 3.3, as well as Chapter 20 of [4].

To avoid confusion, let us briefly review the notation in use. We will use the variable ϑ to refer to the unknown parameter vector, while θ is the AOA of the incoming signal (in degrees). $\psi = \theta\pi/180$ is the AOA (in radians). Whenever a trigonometric function is used, the argument is provided in radians, rather than degrees, to avoid confusion. In MATLAB, one can use the sind, cosd, and tand functions to supply degree inputs, but we stick to radians in order to make the error calculations more consistent and straightforward. Finally, for complex signals, particularly when the phase is unknown, we will rely on ϕ to denote the phase. $[\mathbf{x}]_i$ denotes the ith element of the vector $\mathbf{x}$ and $[\mathbf{X}]_{i,j}$ is the element in the ith row and jth column of the matrix $\mathbf{X}$.

7.1 BEAM PATTERN-BASED DIRECTION FINDING

The simplest form of direction finding is to search for the bearing that maximizes the received signal power. This approach relies on the use of antennas that have a nonuniform response to signals at different directions of arrival.

Let P be the signal power incident on an antenna. The received signal power R out of the antenna is given

$$R = |G(\psi)|^2 P \tag{7.1}$$

where $G(\psi)$ is the complex voltage gain pattern as a function of bearing angle ψ (in radians).

To obtain an estimate, several measurements are taking at different steering angles ψ_i, and the results are processed to determine an estimate. Consider a vector of M samples taking at steering angle ψ_i.

$$\mathbf{x}_i = G(\psi_i - \psi)\mathbf{s} + \mathbf{n}_i \sim \mathcal{CN}\left(G(\psi_i - \psi)\mathbf{s}, \sigma_n^2 \mathbf{I}_M\right) \tag{7.2}$$

In this case, all of the gains and losses aside from the beampattern are subsumed into $\mathbf{s}$. The full data matrix $\mathbf{X}$ is the collection of samples across a set of steering angles.

$$\mathbf{X} = [\mathbf{x}_1, \ldots, \mathbf{x}_N] \tag{7.3}$$

If we stack the data matrix $\mathbf{X}$ into a vector $\mathbf{x}$, it can be described with the distribution

$$\mathbf{x} \sim \mathcal{CN}\left(\mathbf{s} \otimes \mathbf{g}, \sigma_n^2 \mathbf{I}_{MN}\right) \tag{7.4}$$

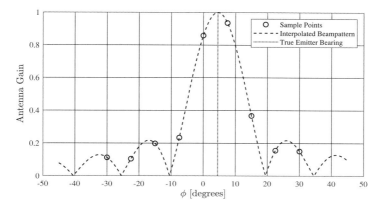

Figure 7.1 Curve fitting the observed gain patterns to obtain an estimate of the AOA.

where $\otimes$ is the Kronecker product,[2] and $\mathbf{g}$ is the vector of amplitude responses at each steering angle

$$\mathbf{g}(\psi) = [G(\psi_1 - \psi), \ldots, G(\psi_M - \psi)]^T \tag{7.5}$$

To obtain an estimate, the simplest approach is to select the steering angle with the maximum average power, noting that $G(\psi_i - \psi)$ has a maximum value when $\psi_i = \psi$.

$$\widehat{\psi} = \arg\max_i \|\mathbf{x}_i\| \tag{7.6}$$

A more sophisticated approach would be to solve for the AOA that most closely fits the observed gain pattern, as shown in Figure 7.1. Let us define the estimated gain pattern at each steering angle θ_i,

$$\widehat{G}_i = \Re\left\{\frac{\mathbf{x}_i^H \mathbf{s}}{\mathbf{s}^H \mathbf{s}}\right\} \tag{7.7}$$

if $\mathbf{s}$ is known, and collect them into the estimated gain vector $\widehat{\mathbf{g}}$. Then, the optimization is given

$$\widehat{\psi} = \arg\min_\psi \|\widehat{\mathbf{g}} - \mathbf{g}(\psi)\|^2. \tag{7.8}$$

2 The Kronecker product of two vectors is the pairwise product of each element in the two inputs. It is equivalent to an outer product, followed by stacking the resultant matrix into a single column or row.

If s is not known, then the estimates are taken from $\mathbf{x}_i^H \mathbf{x}_i$ and normalized by the peak return. Then, the optimization is conducted over two variables, ψ and an unknown scale factor A.

7.1.1 Implementation

The accompanying MATLAB® code has a function that can be used for testing, or as an example of how to build a digital direction finder based on the gain response at a number of pointing angles. The example can be found in the function aoa.directional_df. For both Adcock and rectangular arrays, the gain function (and its derivative, which will be useful in performance analysis) can be generated with the aoa.make_gain_functions module, and then used to generate the test signal. The code below generates a noise-free test signal for N sampling angles equally spaced in angle and M temporal samples at each point, for a signal with $\theta = 5°$ direction of arrival and carrier frequncy given f.

```
s = exp(1i*2*pi*f*(0:M-1)*t_samp);
[g,g_dot] = aoa.make_gain_functions('adcock',d_lam,psi_0);
psi_true = 5*pi/180;
psi_scan = linspace(-pi,pi-2*pi/N,N);
x = g(psi_scan(:)-psi_true) * s;
```

The direction finding algorithm implemented is a multiresolution search that performs a brute force solution to (7.8) first with a step size of 1 degree (on the interval psi_min to psi_max. Then, the interval is shortened, and the step size is reduced by an order of magnitude. This continues until the step size is less than half of psi_res. The algorithm provided searched only over the AOA ψ, and assumes that the scale factor A is known and removed from the received signal, such that only s is passed in. Extension of this algorithm to the more general case would require a second optimization over the possible scale factors A, or generation of a test statistic that is invariant to scale.

7.1.2 Performance

To assess performance, we find a statistical representation of the received signal, in the presence of noise. Let $\mathbf{x}_i$ be the received signal when the array is pointed toward the ith steering position (ψ_i)

$$\mathbf{x}_i = G(\psi_i - \psi)\mathbf{s} + \mathbf{n}, \tag{7.9}$$

Recall from (7.3) that

$$\mathbf{x}_i \sim \mathcal{CN}\left(G(\psi_i - \psi)\mathbf{s}, \sigma_n^2 \mathbf{I}_M\right) \tag{7.10}$$
$$\mathbf{X} = [\mathbf{x}_0, \ldots, \mathbf{x}_{N-1}] \tag{7.11}$$
$$\mathbf{x} \sim \mathcal{CN}\left(\mathbf{s} \otimes \mathbf{g}, \sigma_n^2 \mathbf{I}_{MN}\right) \tag{7.12}$$

where $\mathbf{x}$ is the *vectorized* version of $\mathbf{X}$ (the realignment into a single column vector). We compute the Fisher information matrix according to (6.54)

$$[\mathbf{F}_\vartheta(\mathbf{x})]_{ij} = \mathrm{Tr}\left[\mathbf{C}^{-1}\frac{\partial \mathbf{C}}{\partial \vartheta_i}\mathbf{C}^{-1}\frac{\partial \mathbf{C}}{\partial \vartheta_j}\right] + 2\Re\left\{\left(\frac{\partial \boldsymbol{\mu}}{\partial \vartheta_i}\right)^H \mathbf{C}^{-1}\left(\frac{\partial \boldsymbol{\mu}}{\partial \vartheta_j}\right)\right\} \tag{7.13}$$

with the parameter vector[3]

$$\boldsymbol{\vartheta} = \begin{bmatrix} \mathbf{s}_R^T & \mathbf{s}_I^T & \psi \end{bmatrix}^T \tag{7.14}$$

Assume that the covariance matrix is $\sigma_n^2 \mathbf{I}_{MN}$ and, thus, does not vary θ, so that term falls out of (6.54). Let $\boldsymbol{\mu}_i = G(\psi_i - \psi)\mathbf{s}$. What remains is the partial derivative of $\boldsymbol{\mu}$ with respect to each component of $\boldsymbol{\vartheta}$.[4]

$$\nabla_{\mathbf{s}_R}\boldsymbol{\mu}_i = G(\psi_i - \psi)\mathbf{I} \tag{7.15}$$
$$\nabla_{\mathbf{s}_I}\boldsymbol{\mu}_i = \jmath G(\psi_i - \psi)\mathbf{I} \tag{7.16}$$
$$\nabla_\theta \boldsymbol{\mu}_i = -\mathbf{s}\dot{G}(\psi_i - \psi) \tag{7.17}$$

where $\dot{G}(\psi) = \partial G(\psi)/\partial \psi$ evaluated at $\psi_i - \psi$. Expanding across the different steering angles, the derivative of the full vector $\boldsymbol{\mu}$ is

$$\nabla_\vartheta \boldsymbol{\mu} = \begin{bmatrix} \mathbf{g} \otimes \mathbf{I}_M & \jmath \mathbf{g} \otimes \mathbf{I}_M & -\dot{\mathbf{g}} \otimes \mathbf{s} \end{bmatrix} \tag{7.18}$$

where the vectors $\mathbf{g}$ and $\dot{\mathbf{g}}$ are defined

$$[\mathbf{g}]_i = G(\psi_i - \psi) \tag{7.19}$$
$$[\dot{\mathbf{g}}]_i = \dot{G}(\psi_i - \psi) \tag{7.20}$$

---

[3] Where $\mathbf{s}_R$ and $\mathbf{s}_I$ are the real and imaginary components of $\mathbf{s}$, respectively.
[4] The gradient $\nabla_\vartheta \mathbf{f}(\vartheta)$ is the collection of partial derivatives with respect to each of the elements of ϑ, $\nabla_\vartheta \mathbf{f}(\vartheta) = [\partial \mathbf{f}(\vartheta)/\partial \vartheta_0, \ldots, \partial \mathbf{f}(\vartheta)/\partial \vartheta_{N-1}]$.

Plugging this into the Fisher information matrix (and ignoring terms that have no real component), we have the matrix

$$\mathbf{F}_\vartheta(\mathbf{X}) = \frac{2}{\sigma_n^2} \left[\begin{array}{cc|c} \mathbf{g}^T\mathbf{g}\mathbf{I}_M & \mathbf{0}_{M,M} & -\mathbf{g}^T\dot{\mathbf{g}}\mathbf{s}_R \\ \mathbf{0}_{M,M} & \mathbf{g}^T\mathbf{g}\mathbf{I}_M & -\mathbf{g}^T\dot{\mathbf{g}}\mathbf{s}_I \\ \hline -\mathbf{g}^T\dot{\mathbf{g}}\mathbf{s}_R^T & -\mathbf{g}^T\dot{\mathbf{g}}\mathbf{s}_I^T & \dot{\mathbf{g}}^T\dot{\mathbf{g}}\mathbf{s}^H\mathbf{s} \end{array} \right] \quad (7.21)$$

To invert this, we apply the Matrix inversion lemma discussed in (6.49), where the matrices $\mathbf{A}$, $\mathbf{B}$, $\mathbf{C}$, and $\mathbf{D}$ are delineated by the dashed lines in (7.21).

$$C_\psi \geq \left(\mathbf{D} - \mathbf{CA}^{-1}\mathbf{B} \right)^{-1} \quad (7.22)$$

$$\geq \frac{\sigma_n^2}{2} \left(\dot{\mathbf{g}}^T\dot{\mathbf{g}} E_s - \frac{(\mathbf{g}^T\dot{\mathbf{g}})^2}{\mathbf{g}^T\mathbf{g}} \left(\mathbf{s}_R^T\mathbf{s}_R + \mathbf{s}_I^T\mathbf{s}_I \right) \right)^{-1} \quad (7.23)$$

$$\geq \frac{1}{2M\xi \left(\dot{\mathbf{g}}^T\dot{\mathbf{g}} - \frac{(\dot{\mathbf{g}}^T\mathbf{g})^2}{\mathbf{g}^T\mathbf{g}} \right)} \quad (7.24)$$

noting that $\mathbf{s}^H\mathbf{s} = \mathbf{s}_R^T\mathbf{s}_R + \mathbf{s}_I^T\mathbf{s}_I = E_s$ and $\xi = E_s/M\sigma_n^2$. Since $\theta = 180\psi/\pi$, we can compute the CRLB of θ with (6.52)

$$C_\theta \geq \frac{180^2}{2\pi^2 M\xi \left(\dot{\mathbf{g}}^T\dot{\mathbf{g}} - \frac{(\dot{\mathbf{g}}^T\mathbf{g})^2}{\mathbf{g}^T\mathbf{g}} \right)} \quad (7.25)$$

7.1.2.1 Adcock Antenna Performance

Adcock antennas are a convenient way to approximate a cosine pattern and consist of two omnidirectional antennas combined via subtraction. A diagram is shown in Figure 7.2. The gain pattern is given as [5]

$$G(\psi) = 2G_{\max} \sin\left(\pi \frac{d}{\lambda} \cos(\psi) \right) \quad (7.26)$$

for some peak gain $G_{\max}$. If $d/\lambda \leq .25$, then $G(\psi)$ can be approximated with

$$G(\psi) \approx 2G_{\max} \pi \frac{d}{\lambda} \cos(\psi) \quad (7.27)$$

Direction-Finding Systems 113

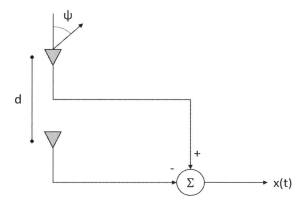

Figure 7.2 Diagram of an Adcock antenna.

The principal advantage of Adcock antennas for direction finding is that they provide a fairly large field of view, but this comes at the expense of poor resolution, particularly when the signal arrives from near broadside (as the signal arriving at the two antennas will be very similar, resulting in significant reduction of SNR). As we will see later in this chapter, they are most frequently used as part of a more complex direction finding system.

For an Adcock antenna with omnidirectional elements ($G_{\max} = 1$)

$$G(\psi) = 2\pi \frac{d}{\lambda} \cos(\psi) \tag{7.28}$$

and the derivative of the gain function is

$$\dot{G}(\psi) = -2\pi \frac{d}{\lambda} \sin(\psi) \tag{7.29}$$

From these, the inner products are written as

$$\mathbf{g}^T \mathbf{g} = \left(2\pi \frac{d}{\lambda}\right)^2 \sum_{i=0}^{2M-1} \cos^2(\psi_i - \psi) \tag{7.30}$$

$$\dot{\mathbf{g}}^T \mathbf{g} = -\left(2\pi \frac{d}{\lambda}\right)^2 \sum_{i=0}^{2M-1} \sin(\psi_i - \psi) \cos(\psi_i - \psi) \tag{7.31}$$

$$\dot{\mathbf{g}}^T \dot{\mathbf{g}} = \left(2\pi \frac{d}{\lambda}\right)^2 \sum_{i=0}^{2M-1} \sin^2(\psi_i - \psi) \tag{7.32}$$

One can then insert these summations direction into (7.24). For brevity, we omit that exercise.

In Figure 7.3, we plot (7.24) for a two-element Adcock antenna with baseline $d = \lambda/4$, as a function of SNR ξ and the number of time samples taken at each antenna position M.[5] In each case, there are $N = 10$ different pointing angles sampled, equally spaced in a circle. We can see that, for high SNR, the Monte Carlo simulation results match closely with the Monte Carlo bound. At low SNR, we see a discrepancy between the CRLB and simulation results, caused by the fact that estimation errors are not truly Gaussian, since there is an upper bound at 180°.

7.1.2.2 Reflector Antenna Performance

A second class of antenna consists of those with a focused beampattern. The specific method of how this is achieved varies greatly among designs, but the two main classes are reflectors (which result in reflected beams that travel in a consistent direction) and arrays (which can create planar wavefronts by radiating a signal from multiple elements). Illustrations of these concepts are provided in Figure 7.4.

Arrays, in particular, have detailed mathematical techniques for direction finding, which we will discuss in Chapter 8. In this section, what matters is only that a directional beam is generated and scanned across the field of view. The smaller beamwidth and higher peak gain of a highly directional antenna can yield significantly improved AOA estimation accuracy, compared with a wide field of view antenna, but their comparatively narrow instantaneous field of view complicates the search for emitters to the side or rear of an antenna.

---

5 In this test, the Monte Carlo simulation results achieved stability when the number of trials exceeded 1,000.

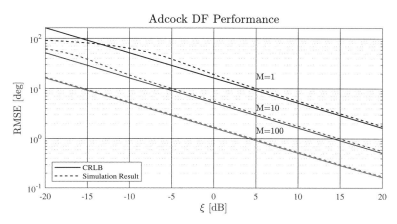

Figure 7.3 Illustration of CRLB performance for an amplitude-based DOA with an Adcock antenna, compared with results from a Monte Carlo trial.

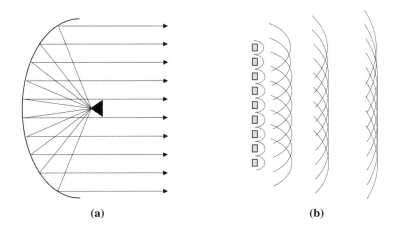

Figure 7.4 Illustration of focused beampatterns with: (a) a properly shaped reflector, and (b) an array of transmitters.

The beampattern for a uniform one-dimensional aperture antenna can be approximated as

$$G(\psi) = G_{\max}\text{sinc}\left(\frac{\psi D}{2\lambda}\right) \quad (7.33)$$

where D is the length of the aperture and λ is the wavelength (in meters), ψ is the angle between the AOA and the pointing angle of the reflector (in radians), and $\text{sinc}(x) = \sin(x)/x$ is the sinc function.[6] To include the effects of elevation angle, a more complex gain pattern is required.[7]

For a generic directional (sinc) beampattern, we have[8]

$$G(\psi) = G_{\max}\text{sinc}\left(\frac{\psi D}{2\lambda}\right) \quad (7.34)$$

$$\dot{G}(\psi) = \begin{cases} 0 & \psi = 0 \\ \dfrac{G_{\max}D}{2\lambda}\dfrac{\left[\frac{\psi D}{2\lambda}\cos\left(\frac{\psi D}{2\lambda}\right) - \sin\left(\frac{\psi D}{2\lambda}\right)\right]}{\left(\frac{\psi D}{2\lambda}\right)^2} & \psi \neq 0 \end{cases} \quad (7.35)$$

From this, the inner products are written as

$$\mathbf{g}^T\mathbf{g} = G_{\max}^2 \sum_{i=0}^{M-1} \text{sinc}^2\left(\frac{D(\psi_i - \psi)}{2\lambda}\right) \quad (7.36)$$

$$\dot{\mathbf{g}}^T\mathbf{g} = \frac{G_{\max}^2 D}{2\lambda} \sum_{i=0}^{M-1} \begin{cases} 0 & \psi_i = \psi \\ \text{sinc}(\tilde{\psi}_i)\dfrac{[\tilde{\psi}_i\cos(\tilde{\psi}_i) - \sin(\tilde{\psi}_i)]}{(\tilde{\psi}_i)^2} & \psi_i \neq \psi \end{cases} \quad (7.37)$$

where, for brevity, $\tilde{\psi}_i = (\psi_i - \psi)D/2\lambda$.

The error estimation prediction, along with results from a Monte Carlo simulation, are shown in Figure 7.5 for a case with $N = 10$ angular samples around a small cue area near the true AOA ($\theta_i \in [-45°, 45°]$, $\theta = 5°$).[9] The aperture has length $D = 5\lambda$, which corresponds to a mainlobe beamwidth of $\approx 11.5°$. As with the Adcock antenna case, we see good agreement between the simulated results and the CRLB, but poor agreement at low SNR, as the error begins to rise above $10°$.

[6] Note that MATLAB's sinc operator is defined slightly differently, as $\text{sinc}(x) = \sin(\pi x)/(\pi x)$.
[7] Interested readers are directed to [6] for detailed discussion of array beampatterns for specialized and generic apertures and array types.
[8] To solve the derivative, we note that $\partial \text{sinc}(x)/\partial x = [x\cos(x) - \sin(x)]/x^2$. Also, the limit as $x \to 0$ is 0.
[9] It is important to ensure that at least one sample is offset from θ by less than the beamwidth of the antenna, such that at least one sample of the beampattern's mainlobe is collected. In this case, $\theta_{3\text{dB}} \approx 11.5°$, so sample spacing of $\theta_i - \theta_{i-1} = 10°$ is sufficient.

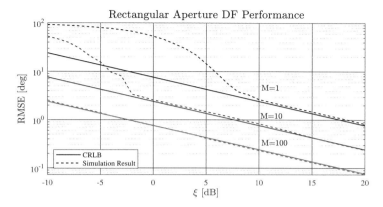

Figure 7.5 Direction of arrival performance as a function of ξ for $N = 11$ samples taken at various offset angles from true target position.

7.2 WATSON-WATT DIRECTION FINDING

Watson-Watt is an amplitude comparison technique, ideally comparing two cosine patterns that are orthogonal to each other, as shown in Figure 7.6, along with the beam patterns for the three channels. This is most simply achieved by constructing two different two-element Adcock antennas, oriented 90° from each other. In this manner, we have two output signals [5]

$$x(t) = V_x \cos(\psi) \cos(\omega t + \pi/2) \quad (7.38)$$
$$y(t) = V_y \sin(\psi) \cos(\omega t + \pi/2) \quad (7.39)$$

where the phase shift $\pi/2$ on the carrier signals is a result of the difference operation in the Adcock antenna. Both signals are typically mixed with a reference

$$r(t) = V_r \cos(\omega t) \quad (7.40)$$

To account for the phase shift, the reference signal is similarly phase-shifted, and then the test signals are filtered with the phase conjugate of the reference signal (or,

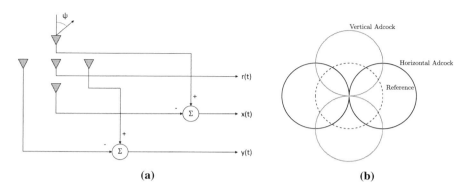

Figure 7.6 Illustration of the (a) system schematic and (b) antenna pattern for a typical Watson-Watt direction-finding system.

equivalently, divided by the reference signal).

$$x = \frac{1}{T}\int_0^T x(t)r^*(t)e^{-j\pi/2}dt = V^2\cos(\psi) \quad (7.41)$$

$$y = \frac{1}{T}\int_0^T y(t)r^*(t)e^{-j\pi/2}dt = V^2\sin(\psi) \quad (7.42)$$

The phase can be computed from the arctan

$$\widehat{\theta} = \tan^{-1}(y/x) \quad (7.43)$$

Note that there is ambiguity in (7.43), to handle this, we make use of an unambiguous implementation:

```
th_hat = atan2(y,x);
```

7.2.1 Implementation

The accompanying MATLAB® code contains an implementation of a digital Watson-Watt receiver, in the aoa.watson_watt_df function. It is called by supplying a reference signal, and the two test (x and y) signals, each as a vector of length M. The code is a straightforward execution of (7.41)-(7.43).

```
r = cos(2*pi*f0*(0:M-1)*t_samp);
x = sin(psi_true)*r;
y = cos(psi_true)*r;
th_est = aoa.watson_watt_df(r,x,y);
```

7.2.2 Performance

The commonly cited rule of thumb for Watson-Watt receivers is that, given sufficient SNR, the angle error is no better than $\approx 2.5°$ RMS [2]. This performance limit is dependent on the realizable tolerances and sensitivities for all of the RF components necessary to build a Watson-Watt DF system and is subject to change with digital implementations. Nevertheless, the analysis below ignores all details of implementation and present a theoretical bound rooted in statistical signal processing.

Consider a digital representation of the Watson-Watt receiver

$$\mathbf{x} = \cos(\psi)\mathbf{s} + \mathbf{n}_x \tag{7.44}$$

$$\mathbf{y} = \sin(\psi)\mathbf{s} + \mathbf{n}_y \tag{7.45}$$

$$\mathbf{r} = \mathbf{s} + \mathbf{n}_r \tag{7.46}$$

where $\mathbf{s}$ is the (real-valued) received signal, and $\mathbf{n}_x$, $\mathbf{n}_y$, and $\mathbf{n}_r$ are all independent Gaussian random variables with variance σ_n^2. If we define a combined data vector $\mathbf{z}$

$$\mathbf{z} = \begin{bmatrix} \mathbf{x}^T & \mathbf{y}^T & \mathbf{r}^T \end{bmatrix}^T \tag{7.47}$$

Note that, since the Watson-Watt system is defined as an amplitude comparison approach, we are restricting the data vectors to be real-valued (rather than complex). As such, the Fisher information matrix is given by (6.53)

$$[\mathbf{F}_\vartheta(\mathbf{x})]_{ij} = \frac{1}{2}\text{Tr}\left[\mathbf{C}^{-1}\frac{\partial \mathbf{C}}{\partial \vartheta_i}\mathbf{C}^{-1}\frac{\partial \mathbf{C}}{\partial \vartheta_j}\right] + \left(\frac{\partial \boldsymbol{\mu}}{\partial \vartheta_i}\right)^T \mathbf{C}^{-1}\left(\frac{\partial \boldsymbol{\mu}}{\partial \vartheta_j}\right) \tag{7.48}$$

We define the parameter vector

$$\boldsymbol{\vartheta} = \begin{bmatrix} \mathbf{s}^T & \psi \end{bmatrix}^T \tag{7.49}$$

and note that $\mathbf{C}$ is independent of the parameter vector $\boldsymbol{\vartheta}$, so the first component of the FIM is zero. The partial derivative of the mean vector with respect to each entry in the parameter vector is given as:

$$\nabla_\vartheta \mu = \begin{bmatrix} \cos(\psi)\mathbf{I}_M & -\sin(\psi)\mathbf{s} \\ \sin(\psi)\mathbf{I}_M & \mathbf{s} \\ \mathbf{I}_M & \mathbf{0}_{M,1} \end{bmatrix} \quad (7.50)$$

Plugging the partial derivative into (7.48) yields the Fisher information matrix:

$$\mathbf{F}_\vartheta(\mathbf{x}) = \frac{1}{\sigma_n^2} \begin{bmatrix} 2\mathbf{I}_M & \mathbf{0}_{M,1} \\ \mathbf{0}_{1,M} & \mathbf{s}^T\mathbf{s} \end{bmatrix} \quad (7.51)$$

Since $\mathbf{F}_\vartheta(\mathbf{x})$ is diagonal, the inverse is trivially computed by inverting the diagonal elements. Thus, bottom right element (which corresponds to the CRLB of ψ) is

$$C_\psi \geq \left[\mathbf{J}_\vartheta(\mathbf{x})^{-1}\right]_{2,2} = \frac{\sigma_n^2}{\mathbf{s}^T\mathbf{s}} \quad (7.52)$$

We define the SNR $\xi = \mathbf{s}^T\mathbf{s}/M\sigma_n^2$, and the CRLB is rewritten as

$$C_\psi \geq \frac{1}{M\xi} \qquad C_\theta \geq \left(\frac{180}{\pi}\right)^2 \frac{1}{M\xi} \quad (7.53)$$

The first critical observation is that performance is now independent of θ. With a single Adcock antenna, there were nulls in the pattern, leading to angles at which the DF system would perform poorly. By combining two Adcock antennas aligned orthogonally, uniform angular accuracy is obtained regardless of the angle of arrival.

Figure 7.7 plots (7.53) as a function of the received SNR (ξ), for several different sample vector sizes M. Note that (7.53) is a bound on the mean squared error, while the $2.5°$ ideal performance limit from [1] is a bound on root mean squared error. To compare them fairly, Figure 7.7 plots the square root of (7.53).

7.3 DOPPLER-BASED DIRECTION FINDING

Another method of direction finding is to exploit the Doppler effect by comparing frequency received at a stationary reference antenna, and a test antenna rotating around the reference, as seen in Figure 7.8, along with the received signal frequency, as the test antenna rotates. At the time of the negative going zero crossing, the angle from the reference to the test antenna is also the AOA of the incoming signal. Alternatively, a series of antennas can be placed in a ring around the reference, and sampled iteratively, as shown in Figure 7.9. For further detail, see [5].

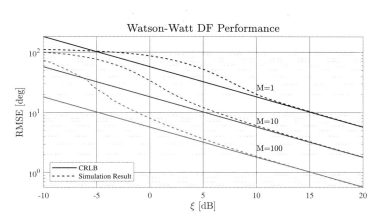

Figure 7.7 Plot of the square root of (7.53) as a function of ξ, for various sample vector lengths M.

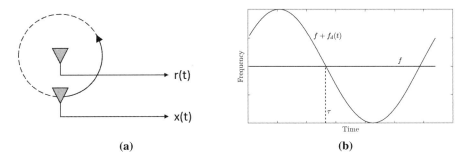

Figure 7.8 A Doppler direction-finding system consists of a test and reference antenna, with the former traveling in a circle around the latter: (a) schematic and (b) relative frequencies.

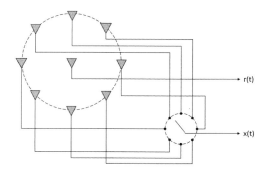

Figure 7.9 A sampled Doppler direction-finding system uses a ring of static test antennas that are switched rapidly to mimic a rotating antenna.

7.3.1 Formulation

We begin with the reference antenna, sampled at times $t_s i$, $i = 0, \ldots, M-1$, to form the vector **r**:

$$[\mathbf{r}]_i = A e^{j\phi_0} e^{j\omega t_s i} + [\mathbf{n}_r]_i \tag{7.54}$$

where ϕ_0 is the unknown starting phase of the signal, A is the signal amplitude (assumed to be constant across the collection interval), $\omega = 2\pi f$ is the signal's central frequency, and t_s is the sampling rate. The noise term $[\mathbf{n}_r]_i$ is assumed to be complex Gaussian with zero mean and variance σ_n^2. Note that this ignores any kind of modulation on the signal.

The rotating antenna's output is similarly given by **x**

$$[\mathbf{x}]_i = A e^{j\phi_0} e^{j\omega t_s i} e^{j\omega \frac{R}{c} \cos(\omega_r t_s i - \psi)} + [\mathbf{n}_x]_i \tag{7.55}$$

where $\omega_r = 2\pi f_r$ is the rotational frequency, R is the radius of the circle around which the antenna rotates, c is the speed of light, and ψ is the AOA of the signal (in radians). The noise term $[\mathbf{n}_x]_i$ is similarly defined.

By comparing the two signals, we have **y**

$$[\mathbf{y}]_i = A^2 e^{j\omega \frac{R}{c} \cos(\omega_r t_s i - \psi)} + [\mathbf{n}_y]_i \tag{7.56}$$

where $[\mathbf{n}_y]_i$ includes not just the product of the noise terms $[\mathbf{n}_x]_i$ and $[\mathbf{n}_r]_i$, but also multiplicative terms with the two signal portions of $[\mathbf{x}]_i$ and $[\mathbf{r}]_i$.[10]

---

10 The distribution of $\mathbf{y}_i$ is given by the *complex double Gaussian* distribution, which described the product of two independent complex Gaussian random variables [7].

Next, we estimate the instantaneous frequency of **y**, and use that to compute the phase shift at each sample $\boldsymbol{\phi}$ and direction of arrival ψ.[11]

$$[\boldsymbol{\phi}]_i = \tan^{-1}\left(\frac{[\mathbf{y}_I]_i}{[\mathbf{y}_R]_i}\right) \approx \omega \frac{R}{c} \cos(\omega_r t_s i - \psi) \tag{7.57}$$

From here, estimation follows the same approach as in Section 7.1; we either select the sample that is closest to a zero-crossing (in the negative direction)

$$\widehat{i} = \arg\min_i |[\boldsymbol{\phi}]_i|, \quad s.t. \ \phi_{i+1} < \phi_i \tag{7.58}$$

$$\widehat{\psi} = w_r t_s \widehat{i} \tag{7.59}$$

or perform a more general optimization

$$\widehat{\psi} = \arg\min_\psi \sum_{i=0}^{N-1} \left|[\boldsymbol{\phi}]_i - \omega \frac{R}{c} \cos(\omega_r t_s i - \psi)\right|^2 \tag{7.60}$$

7.3.2 Implementation

An example implementation, based on a brute force approach to the optimization in (7.60), is given in the provided MATLAB® code, in the function aoa.doppler_df. It is called with the parameters of the Doppler system

```
r = exp(1i*2*pi*f*(0:M-1)*t_s);
x = r .* exp(1i*2*pi*f*R/c*cos(2*pi*fr*(0:M-1)*t_s-psi_true));
psi_est = aoa.doppler_df(r,x,ts,f,R,fr,psi_res,psi_min,psi_max);
```

where r and x are the reference and test signals, respectively; f is the carrier frequency; M is the number of samples; t_s is the sampling period; R is the radius of the Doppler antenna's circular path; c is the speed of light; fr is the Doppler antenna's rotational frequency (in revolutions per second); and psi_true is the true AOA (in radians).

The solver parameters psi_res, psi_min, and psi_max define the desired estimate's resolution and bounds on the region to search. This algorithm is carried out in the same manner as the sample implementation for DF from directional antennas, described in Section 7.1.1; the region of interest is first sampled with 1-degree spacing, the best estimate is determined, and then the process is iteratively repeated with smaller regions of interest and tighter sampling, until the desired sample spacing psi_res is achieved.

---

11 $\mathbf{y}_R$ and $\mathbf{y}_I$ are the real and imaginary components of **y**, respectively.

7.3.3 Performance

The commonly cited rule of thumb for Doppler DF systems is that, given sufficiently high SNR, angle error will be no better than $\approx 2.5°$ RMS. As with the case of Watson-Watt, this is a performance bound based on limitations in installed performance, while the analysis below will consider only the theoretical formulations for a Doppler DF system.

Before we begin, we note that the test and reference data are statistically independent, so we can separate the Fisher information matrix into its components

$$\mathbf{F}_\vartheta(\mathbf{x},\mathbf{r}) = \mathbf{F}_\vartheta(\mathbf{x})\mathbf{F}_\vartheta(\mathbf{r}|\mathbf{x}) = \mathbf{F}_\vartheta(\mathbf{x}) + \mathbf{F}_\vartheta(\mathbf{r}) \tag{7.61}$$

with unknown parameter vector ϑ defined

$$\vartheta = \begin{bmatrix} A & \phi_0 & \omega & \psi \end{bmatrix} \tag{7.62}$$

We begin with the reference data, which is independent and identically distributed according to

$$[\mathbf{r}]_i \sim \mathcal{CN}\left(Ae^{j\phi_0}e^{j\omega t_s i}, \sigma_n^2\right) \tag{7.63}$$

For simplicity, we define the signal vector

$$[\mathbf{s}_r]_i = e^{j\omega t_s i} \tag{7.64}$$

The derivative of $\mathbf{r}$ with respect to the parameter vector is given as

$$\nabla_\vartheta \mu_r = e^{j\phi_0} \begin{bmatrix} \mathbf{s}_r & jA\mathbf{s}_r & jA\dot{\mathbf{s}}_r & \mathbf{0}_{M,1} \end{bmatrix} \tag{7.65}$$

where $\dot{\mathbf{s}}_r$ is defined as

$$[\dot{\mathbf{s}}_r]_i = it_s [\mathbf{s}_r]_i \tag{7.66}$$

Note that the final column in the gradient matrix is zero; this is because the reference signal has no information on the value of ψ; for that we need the test signal $\mathbf{x}$. From this, we can compute the Fisher information matrix of $\mathbf{r}$ using (6.54)

$$\mathbf{F}_\vartheta(\mathbf{r}) = \frac{2}{\sigma_n^2} \Re \left\{ \begin{bmatrix} \mathbf{s}_r^H \mathbf{s}_r & jA\mathbf{s}_r^H \mathbf{s}_r & jA\mathbf{s}_r^H \dot{\mathbf{s}}_r & 0 \\ -jA\mathbf{s}_r^H \mathbf{s}_r & A^2\mathbf{s}_r^H \mathbf{s}_r & A^2\mathbf{s}_r^H \dot{\mathbf{s}}_r & 0 \\ -jA\dot{\mathbf{s}}_r^H \mathbf{s}_r & A^2\dot{\mathbf{s}}_r^H \mathbf{s}_r & A^2\dot{\mathbf{s}}_r^H \dot{\mathbf{s}}_r & 0 \\ 0 & 0 & 0 & 0 \end{bmatrix} \right\} \tag{7.67}$$

At this point, we make a few observations. $\mathbf{s}_r^H \mathbf{s}_r = M$, $\mathbf{s}_r^H \dot{\mathbf{s}}_r = t_s M(M-1)/2$, and $\dot{\mathbf{s}}_r^H \dot{\mathbf{s}}_r = t_s^2 M(M-1)(2M-1)/6$. Also, we define the SNR $\xi = A^2/\sigma_n^2$.[12] Using these solutions, we can simplify the $\mathbf{F}_\theta(\mathbf{r})$ as

$$\mathbf{F}_\vartheta(\mathbf{r}) = 2M\xi \begin{bmatrix} 1/A^2 & 0 & 0 & 0 \\ 0 & 1 & t_s \frac{M-1}{2} & 0 \\ 0 & t_s \frac{M-1}{2} & t_s^2 \frac{(M-1)(2M-1)}{6} & 0 \\ 0 & 0 & 0 & 0 \end{bmatrix} \quad (7.68)$$

Next, we compute the FIM of $\mathbf{x}$ in a similar manner. Note that each element is IID with

$$[\mathbf{x}]_i \sim \mathcal{CN}\left(Ae^{\jmath\phi_0}[\mathbf{s}_x]_i, \sigma_n^2\right) \quad (7.69)$$

$$[\mathbf{s}_x]_i = e^{\jmath\omega t_s i} e^{\jmath\omega \frac{R}{c} \cos(\omega_r t_s i - \psi)} \quad (7.70)$$

The gradient vector of the expectation $\boldsymbol{\mu}_x$ is given as

$$\nabla_\vartheta \boldsymbol{\mu}_x = e^{\jmath\phi_0} \begin{bmatrix} \mathbf{s}_x & \jmath A \mathbf{s}_x & \jmath A \dot{\mathbf{s}}_x & \jmath A \ddot{\mathbf{s}}_x \end{bmatrix} \quad (7.71)$$

$$[\dot{\mathbf{s}}_x]_i \triangleq \left(t_s i + \frac{R}{c}\cos(\omega_r t_s i - \psi)\right)[\mathbf{s}_x]_i \quad (7.72)$$

$$[\ddot{\mathbf{s}}_x]_i \triangleq \omega \frac{R}{c}\sin(\omega_r t_s i - \psi)[\mathbf{s}_x]_i \quad (7.73)$$

Using these gradients, the FIM of $\mathbf{x}$ is written as

$$\mathbf{F}_\vartheta(\mathbf{x}) = \frac{2}{\sigma_n^2}\Re\left\{\begin{bmatrix} \mathbf{s}_x^H \mathbf{s}_x & \jmath A \mathbf{s}_x^H \mathbf{s}_x & \jmath A \mathbf{s}_x^H \dot{\mathbf{s}}_x & \jmath A \mathbf{s}_x^H \ddot{\mathbf{s}}_x \\ -\jmath A \mathbf{s}_x^H \mathbf{s}_x & A^2 \mathbf{s}_x^H \mathbf{s}_x & A^2 \mathbf{s}_x^H \dot{\mathbf{s}}_x & A^2 \mathbf{s}_x^H \ddot{\mathbf{s}}_x \\ -\jmath A \dot{\mathbf{s}}_x^H \mathbf{s}_x & A^2 \dot{\mathbf{s}}_x^H \mathbf{s}_x & A^2 \dot{\mathbf{s}}_x^H \dot{\mathbf{s}}_x & A^2 \dot{\mathbf{s}}_x^H \ddot{\mathbf{s}}_x \\ -\jmath A \ddot{\mathbf{s}}_x^H \mathbf{s}_x & A^2 \ddot{\mathbf{s}}_x^H \mathbf{s}_x & A^2 \ddot{\mathbf{s}}_x^H \dot{\mathbf{s}}_x & A^2 \ddot{\mathbf{s}}_x^H \ddot{\mathbf{s}}_x \end{bmatrix}\right\} \quad (7.74)$$

In this case, only one simplification is straightforward. $\mathbf{s}_x^H \mathbf{s}_x = M$. For the remaining terms, we can simplify to a summation over $i = 0, \ldots, M-1$, and

---

[12] There is a slight discrepancy in the definition of ξ here, with respect to all other uses in this text. This is caused by the way we define $\mathbf{s}$ in (7.64) to have constant amplitude. The signal power, caused by A in this formulation, is kept separate for the CRLB analysis and, thus, must be considered explicitly when defining ξ.

conclude that each is real.

$$\mathbf{s}^H\dot{\mathbf{s}} = \sum_{i=0}^{M-1}\left(t_si + \frac{R}{c}\cos(\omega_r t_s i - \psi)\right) \quad (7.75)$$

$$\mathbf{s}^H\ddot{\mathbf{s}} = \sum_{i=0}^{M-1}\omega\frac{R}{c}\sin(\omega_r t_s i - \psi) \quad (7.76)$$

$$\dot{\mathbf{s}}^H\dot{\mathbf{s}} = \sum_{i=0}^{M-1}\left(t_s i + \frac{R}{c}\cos(\omega_r t_s i - \psi)\right)^2 \quad (7.77)$$

$$\dot{\mathbf{s}}^H\ddot{\mathbf{s}} = \sum_{i=0}^{M-1}\left(t_s i + \frac{R}{c}\cos(\omega_r t_s i - \psi)\right)\omega\frac{R}{c}\sin(\omega_r t_s i - \psi) \quad (7.78)$$

$$\ddot{\mathbf{s}}^H\ddot{\mathbf{s}} = \sum_{i=0}^{M-1}\left(\omega\frac{R}{c}\sin(\omega_r t_s i - \psi)\right)^2 \quad (7.79)$$

Removing the imaginary terms, the FIM of **x** is given as

$$\mathbf{F}_\vartheta(\mathbf{x}) = 2M\xi \begin{bmatrix} \frac{1}{A^2} & 0 & 0 & 0 \\ 0 & 1 & \frac{1}{M}\mathbf{s}_x^H\dot{\mathbf{s}}_x & \frac{1}{M}\mathbf{s}_x^H\ddot{\mathbf{s}}_x \\ 0 & \frac{1}{M}\dot{\mathbf{s}}_x^H\mathbf{s}_x & \frac{1}{M}\dot{\mathbf{s}}_x^H\dot{\mathbf{s}}_x & \frac{1}{M}\dot{\mathbf{s}}_x^H\ddot{\mathbf{s}}_x \\ 0 & \frac{1}{M}\ddot{\mathbf{s}}_x^H\mathbf{s}_x & \frac{1}{M}\ddot{\mathbf{s}}_x^H\dot{\mathbf{s}}_x & \frac{1}{M}\ddot{\mathbf{s}}_x^H\ddot{\mathbf{s}}_x \end{bmatrix} \quad (7.80)$$

To compute the CRLB, one must combine (7.68) and (7.80), and then numerically invert the matrix. The error variance of $\hat{\psi}$ will be bounded by the bottom-right element of the inverse

$$C_\psi \geq \left[(\mathbf{F}_\vartheta(\mathbf{x}) + \mathbf{F}_\vartheta(\mathbf{r}))^{-1}\right]_{4,4} \quad (7.81)$$

Once that is computed, the variance in degrees is a straightforward multiplication

$$C_\theta = \left(\frac{180}{\pi}\right)^2 C_\psi \quad (7.82)$$

Figure 7.10 plots the CRLB for a Doppler DF receiver with M samples taken around the unit sphere. The Monte Carlo simulation was conducted for 10^7 random trials, and shows good agreement with the CRLB prediction at high SNR. Performance rapidly collapsed for SNR below $\xi \approx 10$ dB, and then saturates near $\sigma_\theta \approx 100°$. This saturation effect is consistent with the other DF methods discussed in this chapter.

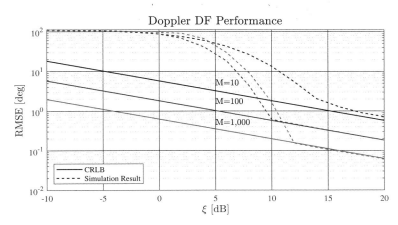

Figure 7.10 Direction-finding performance for a Doppler receiver with M samples collected over the course of one revolution.

7.4 PHASE INTERFEROMETRY

Phase interferometry collects coherent complex measurements at two or more antenna positions, and carefully compares the phase information between them. The mathematics of phase interferometry are equivalent to that of more general phased arrays, but in this section, we consider the case of two antennas, separated by some distance d. The more general case will be discussed in Chapter 8. Phase interferometry is discussed in [8–10], and Chapter 3 of [11]. It is also derived in the context of source localization from a single satellite in Chapter 3 of [12].

The beam pattern of a two-element interferometer is given by

$$G(\psi) = G_{\max} \left(1 + e^{j2\pi \frac{d}{\lambda}(\sin(\psi) - \sin(\psi_s))}\right) \quad (7.83)$$

where ψ is the AOA of the incoming signal and ψ_s is the steering vector of the received beam former, both in radians. Figure 7.12 plots (7.83) for various antenna spacings (in units of wavelengths). This pattern is equal to $2G_{\max}$ when $\psi = \psi_s$, as well as any other angles where $d(\sin(\psi) - \sin(\psi_s))/\lambda$ is a multiple of .5.

Note that, for $d/\lambda \leq 1/2$, there is a single lobe, albeit with a large beamwidth. This creates an unambiguous measurement of the signal's AOA, but at low accuracy. As d/λ increases, the width of the mainlobe decreases, but ambiguous lobes (called *grating lobes*) emerge, presenting ambiguities in regular intervals. In these cases, it

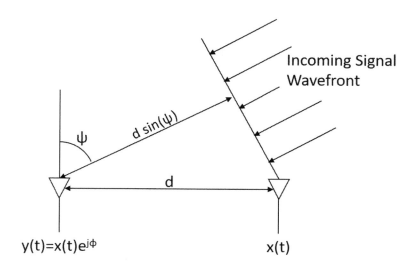

Figure 7.11 Depiction of a phase interferometer, with two sensors measuring the complex phase ϕ between them, and its relation to the direction of arrival for a planar wavefront.

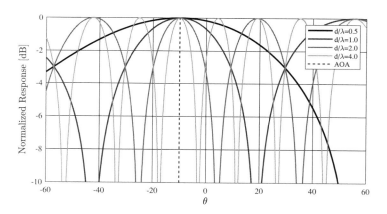

Figure 7.12 Beam patterns for phase interferometers with multiple baselines.

is impossible to determine which of the grating lobes represents the true AOA of an incoming signal.

Taking the phase relation in Figure 7.11, we can construct a digital representation of the sampled data in vector form:

$$\mathbf{x}_1 = \mathbf{s} + \mathbf{n}_1 \qquad (7.84)$$

$$\mathbf{x}_2 = \alpha \mathbf{s} e^{\jmath\phi} + \mathbf{n}_2 \qquad (7.85)$$

where $\mathbf{n}_1, \mathbf{n}_2 \sim \mathcal{CN}\left(0, \sigma_n^2 \mathbf{I}_M\right)$ and the phase shift ϕ is given as

$$\phi = 2\pi \frac{d}{\lambda} \sin(\psi) \qquad (7.86)$$

It can be inferred from (7.85) that the phase of the received signal must be measured. In order to remove the nuisance parameter $\mathbf{s}$, we correlate the two received signals. This is represented as simply the inner product.

$$y = \mathbf{x}_1^H \mathbf{x}_2 = \alpha e^{\jmath\phi} \mathbf{s}^H \mathbf{s} + \underbrace{\mathbf{s}^H \mathbf{n}_2 + \alpha e^{\jmath\phi} \mathbf{n}_1^H \mathbf{s}}_{cross-terms} + \underbrace{\mathbf{n}_1^H \mathbf{n}_2}_{noise}. \qquad (7.87)$$

The estimate, is thus given by computing the phase of y[13]

$$\widehat{\phi} = \tan^{-1}\left(\frac{y_I}{y_R}\right) \qquad (7.88)$$

and then translating to an AOA

$$\widehat{\psi} = \sin^{-1}\left(\frac{\lambda \widehat{\phi}}{2\pi d}\right) \qquad (7.89)$$

7.4.1 Implementation

The enclosed MATLAB® code contains a straightforward implementation of (7.89), and can be found in the `aoa.interf_df` function.

```
x1 = exp(1i*2*pi*f*(0:M-1)*t_s);
x2 = alpha * x1 .* exp(1i*2*pi*d_lam*sin(psi_true));
psi_est = aoa.interf_df(x1,x2,d_lam);
```

---

[13] y_R and y_I are the real and imaginary components of y, respectively.

where f is the carrier frequency, M is the numbers of samples, t_s is the sampling period, alpha is the relative scale factor between the two signals, d_lam is the spacing between the antennas (in wavelengths), and psi_true is the true AOA (in radians).

7.4.2 Performance

Recall from (6.54) that the Fisher information matrix for a complex Gaussian random variable with mean μ, covariance matrix $\mathbf{C}$ and real parameter vector ϑ is:

$$[\mathbf{F}_\vartheta(\mathbf{x})]_{ij} = \text{Tr}\left[\mathbf{C}^{-1}\frac{\partial \mathbf{C}}{\partial \vartheta_i}\mathbf{C}^{-1}\frac{\partial \mathbf{C}}{\partial \vartheta_j}\right] + 2\Re\left\{\left(\frac{\partial \mu}{\partial \vartheta_i}\right)^H \mathbf{C}^{-1}\left(\frac{\partial \mu}{\partial \vartheta_j}\right)\right\} \quad (7.90)$$

We will apply this result to the combined data vector

$$\mathbf{x} = \begin{bmatrix} \mathbf{x}_1 \\ \mathbf{x}_2 \end{bmatrix} \sim \mathcal{CN}\left(\begin{bmatrix} \mathbf{s} \\ \alpha e^{\jmath\phi}\mathbf{s} \end{bmatrix}, \begin{bmatrix} \sigma_1^2\mathbf{I}_M & \mathbf{0}_M \\ \mathbf{0}_M & \sigma_2^2\mathbf{I}_M \end{bmatrix}\right) \quad (7.91)$$

with the real parameter vector

$$\vartheta = \begin{bmatrix} \mathbf{s}_R^T & \mathbf{s}_I^T & \alpha & \phi \end{bmatrix}^T \quad (7.92)$$

First, we note that the covariance matrix $\mathbf{C}$ does not vary with any of the parameters in ϑ. Thus, $\partial \mathbf{C}/\partial \vartheta_i = 0$, and the first term drops out. The second term can be solved by first computing the gradient for all elements of ϑ [13]

$$\nabla_\vartheta \mu = \begin{bmatrix} \mathbf{I}_M & \jmath\mathbf{I}_M & \mathbf{0}_{M,1} & \mathbf{0}_{M,1} \\ \alpha e^{\jmath\phi}\mathbf{I}_M & \jmath\alpha e^{\jmath\phi}\mathbf{I}_M & e^{\jmath\phi}\mathbf{s} & \jmath\alpha e^{\jmath\phi}\mathbf{s} \end{bmatrix} \quad (7.93)$$

From (7.93) and (7.90) we can construct the Fisher information matrix

$$\mathbf{F}_\vartheta(\mathbf{x}) = 2\begin{bmatrix} \left(\frac{1}{\sigma_1^2}+\frac{\alpha^2}{\sigma_2^2}\right)\mathbf{I}_M & \mathbf{0}_M & \frac{\alpha}{\sigma_2^2}\mathbf{s}_R & -\frac{\alpha^2}{\sigma_2^2}\mathbf{s}_I \\ \mathbf{0}_M & \left(\frac{1}{\sigma_1^2}+\frac{\alpha^2}{\sigma_2^2}\right)\mathbf{I}_M & \frac{\alpha}{\sigma_2^2}\mathbf{s}_I & \frac{\alpha^2}{\sigma_2^2}\mathbf{s}_R \\ \frac{\alpha}{\sigma_2^2}\mathbf{s}_R^T & \frac{\alpha}{\sigma_2^2}\mathbf{s}_I^T & \frac{1}{\sigma_2^2}\mathbf{s}^H\mathbf{s} & 0 \\ -\frac{\alpha^2}{\sigma_2^2}\mathbf{s}_I^T & \frac{\alpha^2}{\sigma_2^2}\mathbf{s}_R^T & 0 & \frac{\alpha^2}{\sigma_2^2}\mathbf{s}^H\mathbf{s} \end{bmatrix} \quad (7.94)$$

Next, we partition the Fisher information matrix according to the dotted lines in (7.94), and apply the matrix inversion lemma from Section 6.4.2.1, resulting in the CRLB for both the scale factor α and phase difference ϕ based on (6.49)[14]

$$\mathbf{C}_{\alpha,\phi} \geq \frac{1}{2}\left(\mathbf{D} - \mathbf{C}\mathbf{A}^{-1}\mathbf{B}\right)^{-1} \qquad (7.95)$$

The evaluation of (7.95) is left to the reader as an exercise, but the result is

$$\mathbf{C}_{\alpha,\phi} \geq \frac{\sigma_1^2 + \frac{\sigma_2^2}{\alpha^2}}{2E_s}\begin{bmatrix} \alpha^2 & 0 \\ 0 & 1 \end{bmatrix} \qquad (7.96)$$

where, $\mathbf{s}^H\mathbf{s} = E_s$. The lower right element of this matrix can be extracted to yield the CRLB of the parameter ϕ when $\mathbf{s}$ and α are unknown nuisance parameters:

$$C_\phi \geq \frac{\sigma_1^2 + \frac{\sigma_2^2}{\alpha^2}}{2E_s} \qquad (7.97)$$

We define the SNR values $\xi_1 = E_s/M\sigma_1^2$ and $\xi_2 = \alpha^2 E_s/M\sigma_2^2$. The effective SNR is then defined as

$$\xi_{\text{eff}} = \left(\xi_1^{-1} + \xi_2^{-1}\right)^{-1} = \frac{E_s/M}{\sigma_1^2 + \frac{\sigma_2^2}{\alpha^2}} \qquad (7.98)$$

and we can rewrite the CRLB in terms of the effective SNR

$$C_\phi \geq \frac{1}{2M\xi_{\text{eff}}} \qquad (7.99)$$

This, however, is the accuracy in determining the phase term ϕ. To convert this error to AOA, we reference the CRLB of a function, given in (6.52). Since both ϕ and ψ are scalars, the forward gradient $\mathbf{G}$ is defined simply as

$$G = \frac{\partial \phi}{\partial \psi} = \frac{2\pi d \cos(\psi)}{\lambda} \qquad (7.100)$$

Thus, the CRLB with respect to AOA is

$$C_\psi \geq C_\phi G^{-2} = \frac{1}{2M\xi_{\text{eff}}}\left(\frac{\lambda}{2\pi d \cos(\psi)}\right)^2 \qquad (7.101)$$

$$C_\theta \geq \left(\frac{180}{\pi}\right)^2 C_\psi \qquad (7.102)$$

---

[14] Even though α is a nuisance parameter, we partition the matrix into four 2x2 squares since that leaves $\mathbf{A}$ and $\mathbf{D}$ as diagonal matrices, simplifying their inversions. After inversion of the matrix, we can isolate the single parameter of interest, ϕ.

Figure 7.13 Graphic of a four-element interferometer with baselines d, 2d, and 4d.

7.4.3 Resolving Ambiguities with Multiple Baselines

Recall from Figure 7.12 that, for $d/\lambda > 1/2$, ambiguities are introduced into the system. This can be resolved in two ways. The first is to use directional antennas as inputs. If the field of view of the inputs is $\pm\theta_{\max}$, then the interferometer baseline can be as large as

$$\frac{d}{\lambda} \leq \frac{1}{2\sin(\theta_{\max})} \tag{7.103}$$

without introducing ambiguities, since any ambiguities that occur are outside the field of view of the antennas. For example, a directional antenna with a 3-dB beamwidth of $\theta_{\text{bw}} = 10°$ can safely be assumed to return very little energy from the region outside $\pm 10°$. Using (7.103), the largest possible baseline to avoid ambiguities is 2.88 wavelengths.

The second principal approach is to implement a multiple baseline interferometer, as shown in Figure 7.13. By placing antennas at regular intervals, it is possible to use the shorter baseline interferometers to isolate which of the beams from a large baseline interferometer corresponds to the true AOA. More formally, modulo arithmetic can be used to isolate the proper solution.

7.5 PERFORMANCE COMPARISON

In this section, we will consider the performance of the various DF receivers discussed in this chapter for estimation of the AOA of a collision-avoidance radar signal, originally described in Table 3.1. Each example's performance will be plotted in a single figure, to enable easier comparison. The parameters for the common transmitter in each example are given in Table 7.1a, and the set of parameters for each receiver are given in Table 7.1b.

Table 7.1
Collision Avoidance Radar Example Parameters for Examples in Section 7.5

(a) CA Radar Transmitter

Parameter	Value
P_t	35 mW
G_t (mainlobe)	34 dBi
B_s	31.3 MHz
f_0	35 GHz
L_t	0 dB

(b) DF Receivers

Parameter	Rectangular	Adcock	Watson-Watt	Doppler	Interferometer
d/λ	5	.25	.25	–	.5
R	–	–	–	$\lambda/2$	–
f_R	–	–	–	$1/T_s$	–
L_r			2 dB		
F_n			4 dB		
B_n			40 MHz		
T			10 μ s		

Example 7.1 Direction Finding with an Adcock Antenna

Consider the detection of the transmitter in Table 7.1a with an Adcock antenna system, as described in Table 7.1b. If the true AOA is $\theta = 10°$, and the Adcock antenna is steered to three angles for testing, $-15°$, $0°$, and $15°$, and dwells at each angle for $T = 10~\mu s$, at what range will the angle estimation error (RMSE) be $\sigma_\theta = 1°$?

We begin by computing the SNR. Note that the SNR referenced above in the performance calculations is that of an omnidirectional antenna; the gain from the Adcock antenna is accounted for in the g and ġ terms. For simplicity, we assume free space path loss, regardless of standoff distance.

```
R = 100:100:50e3;
S = (10*log10(Pt) + Gt - Lt) + 20*log10(lambda./(4*pi*R)) - Lr;
N = 10*log10(k*T*B_n) + F_n;
snr_db = (S-N);
snr_lin = 10.^(snr_db/10);
```

Next, we determine how many samples are taken at each steering angle, based on the noise bandwidth of the receiver, and length T of each collection

```
> M = 1 + floor(T * Br)
```

This results in $M = 400$. Finally, we compute the gain vectors, utilizing the provided function, and note that the returned gain vectors are function handles that accept the difference between the steering and gain vectors, in radians.

```
[g,g_dot] = aoa.make_gain_functions('Adcock',.25,0);
psi_scan = [-15,0,15]*pi/180;
psi_true = 10*pi/180;
g_v = g(psi_scan(:)-psi_true);
gg_v = g_dot(psi_scan(:)-psi_true);
```

All that remains is to evaluate the CRLB

```
crlb_psi = 1./(2*M*snr_lin*((gg_v'*gg_v) - (gg_v'*g_v)^2/(g_v'*g_v)));
```

Alternatively, you take the function handles g and g_dot and call the built in CRLB function

```
crlb_psi = aoa.directional_crlb(snr_db,M,g,g_dot,psi_scan,psi_true)
```

Conversion from the CRLB of ψ to the RMSE of θ is computed via square root and radian to degree conversion

Direction-Finding Systems

```
rmse_theta = (180/pi)*sqrt(crlb_psi);
```

Figure 7.14 plots the RMSE in units of degrees, as a function of the distance between transmitter and receiver, assuming free space path loss (along with several other receivers defined in later example problems). From this example, we can see that the Adcock antenna DF achieves $\sigma_\theta \leq 1°$ at a standoff range of roughly 1.6 km. As in Chapter 3, this problem could easily be inverted, to compute the acceptable standoff range for a required SINR directly, rather than via searching the figure. This is left as an exercise for the reader.

Results are also plotted in Figure 7.14 for a rectangular aperture with $D = 5\lambda$ and the same sample points ψ_i. In this case, the DF system achieves $\sigma_\theta \leq 1°$ at $R \leq 10$ km. However, this result is not robust. Recall that, for $D = 5\lambda$, the beamwidth of the rectangular aperture is $\approx 11.5°$. Since the beamwidth is less than the spacing between sample points, the signal return could fall into a null between two of the sampled angles. To properly use such a directional aperture, the angular space should be sampled more densely.

Example 7.2 Watson-Watt Direction Finding

Repeat Example 7.1, but with a Watson-Watt receiver instead of a single Adcock antenna pair. At what range does the angle estimation error reach $1°$?

The calculations for ξ and M are unchanged. In this case, the CRLB can then be plotted directly from ξ and M.

```
crlb_psi = 1./(M*xi_lin);
rmse_theta = (180/pi)*sqrt(crlb_psi);
```

Alternatively, one can solve for the required SNR, and use the methods in Chapter 3 to back out the range at which that SNR is achieved

```
rmse = 10;
crlb_psi = (pi*rmse/180)^2

snr_max = 1./(M*crlb_psi)
```

The results are plotted for this example in Figure 7.14. The desired accuracy of $1°$ is achieved with a Watson-Watt receiver when $R \leq 3$ km. This is better than the single-channel Adcock antenna, but not as accurate as the high-gain rectangular aperture. Of course, the benefit of the Watson-Watt receiver is that there is instantaneous coverage at all azimuth angles, whereas the others must be scanned mechanically to scan the area of regard.

Example 7.3 Doppler-Based Direction Finding Example

Repeat Example 7.1, but with a Doppler receiver with a radius of $\lambda/2$, and timed to execute a single rotation over the course of a single pulse. At what range does the angle estimation error reach $1°$?

We begin with the SNR vector computed before, and define the parameters necessary for calculation of the CRLB

```
ts = 1/(2*f);
R_dop = lambda/2;
fr = 1/T_s;
crlb_psi_doppler = aoa.doppler_crlb(snr_db,M,1,ts,f0,R_dop,fr,psi_true);
rmse_th_doppler = (180/pi)*sqrt(crlb_psi_doppler);
```

where `psi_true` is the true AOA. This can be plotted and searched to find the point where `rmse_th_doppler` is equal to $1°$. Unfortunately, due to the complexity of the CRLB, it can't be easily computed analytically. The results, along with the prior examples, are plotted in Figure 7.14, and the answer is that the Doppler DF system, as specified, achieves $1°$ accuracy when $R \leq 1$ km.

Example 7.4 Interferometer DF Example

Repeat Example 7.1, but with a two-channel interferometer, with a spacing of $d = \lambda/2$. Assume that $\alpha = 1$, and therefore that both receivers have the same SNR ($\xi_{\text{eff}} = .5\xi_1 = .5\xi_2$). At what range does the angle estimation error reach $1°$? What about if $d = 2\lambda$?

To begin, we invert the CRLB equation and compute the required SNR

```
max_err = (pi/180)^2;
snr_min = 2*(1./(2*M*crlb_max)).*(1./(2*pi*d_lam*cos(psi_true)));
```

The result of this is that the required SNR is 0.6615 (-1.7947 dB). We then apply techniques from Chapter 2 to compute the maximum range at which that SNR is achieved. For simplicity, we evaluate SNR at multiple ranges, and find the value that is closest to the desired range. In the earlier iterations of this example, we computed `snr_db`

```
[~,idx] = min(abs(snr_db - 10*log10(snr_min_lin)));
```

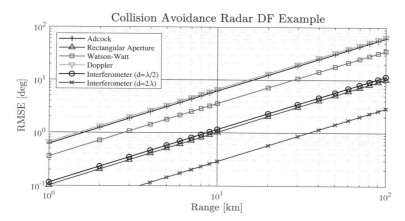

Figure 7.14 Direction-finding RMSE as a function of standoff range for Examples 7.1–7.4

```
R_interf = R(idx)
```

From this, max range is 8.65 km. If we repeat for $d/\lambda = 2$, then the result is 34.6 km. We plot the CRLB for an interferometer in Figure 7.14. For the narrow baseline ($d = \lambda/2$), the figure suggests a $1°$ error is achieved for roughly $R \leq 8.5$ km, while the large baseline ($d = 2\lambda$) achieves that same error at a standoff of $R \leq 35$ km; these numbers match the values calculated explicitly. This example illustrates the power of increasing the baseline d/λ.

7.6 MONOPULSE DIRECTION FINDING

Monopulse direction finding is achieved by forming two beams from a pair of elements, one from their sum and one from their difference, as shown in Figure 7.15. The sum, difference, and ratio patterns are shown in Figure 7.15(b), as a function of the normalized angle (θ/θ_{bw}). The sum beam shows a peak at boresight, while the difference beam has its peaks at ± 0.5 beamwidths. From this, the ratio (plotted as a dashed line) is shown to be linear between ± 0.5 beamwidths.

The heart of monopulse processing is the fact that the ratio of the sum and difference beams is roughly linear in the region around $\theta = 0$. Because of this, if the slope is known, then the output ratio can be simply scaled to yield an estimate of the AOA. The derivation in this section relies heavily on [10], which includes not

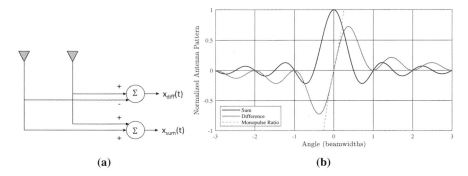

Figure 7.15 Monopulse processing computes the sum and difference beams from a pair of antennas to determine the AOA: (a) architecture, (b) antenna patterns.

only derivations of the relevant equations, as shown here, but also discusses issues of implementation (both hardware and processing algorithms), and performance measures, such as the ability to resolve two closely spaced signals, and careful treatment of angle errors.

Define the two input signals as $x_1(t)$ and $x_t(t)$. The sum and difference beams are given as:

$$x_s(t) = \frac{1}{\sqrt{2}}(x_1(t) + x_2(t)) \tag{7.104}$$

$$x_d(t) = \frac{1}{\sqrt{2}}(x_1(t) - x_2(t)) \tag{7.105}$$

The ratio of the sum and difference beams is then computed

$$r(t) = \frac{x_d(t)}{x_s(t)} \approx k_m \overline{\theta} \tag{7.106}$$

where k_m is defined as the normalized monopulse slope, and $\overline{\theta}$ is the normalized angle, defined as the AOA in beamwidths, rather than radians or degrees, $\overline{\theta} = \theta/\theta_{\text{bw}}$.

From [10], [equation (5.9)], the normalized monopulse slope is given

$$k_m = \frac{\theta_{\text{bw}}}{\sqrt{G_m}} \frac{\partial}{\partial \theta} d\left(\overline{\theta}\right)\bigg|_{\overline{\theta}=0} = \frac{\theta_{\text{bw}} K}{\sqrt{\eta_a}}, \tag{7.107}$$

Table 7.2
Some Typical Monopulse Parameters

Parameter	Ideal Antenna	Cosine Taper
Monopulse slope (K)	1.814	1.334
Beamwidth (θ_{bw})	.886 λ/d	1.189 λ/d
Aperture efficiency (η_a)	1	.657
Monopulse slope (k_m)	1.606	1.957

Note: d is the spacing between the antennas. Taken from [10], Table 5.1.

where $d\left(\overline{\theta}\right)$ is the beampattern of the difference signal, in units of normalized angle, and that derivative is evaluated at boresight. The result depends on the difference signal slope K, and aperture efficiency η_a. The values of K and η_a will depend on the apertures used to feed into the monopulse processor. Table 7.2 shows a few representative values.

Once the normalized monopulse slope k_m is computed (or looked up), then the AOA estimate is computed from the ratio

$$\widehat{\theta}(t) = \frac{\theta_{bw}}{k_m} r(t) \qquad (7.108)$$

Due to the complexities of implementing a monopulse angle of arrival system, and the sheer variety of algorithms in existence, we do not provide a sample implementation. Instead, interested readers are referred to [10] for details of how to construct a monopulse receiver, and the variety of receiver types available, specifically Chapter 5 for stand-alone monopulse receivers, Chapter 7 for monopulse processing in the context of an array, and Chapter 8 for processing algorithms.

7.6.1 Performance

The error signal for a monopulse processor is very complex. For a careful treatment, see Chapter 10 of [10]. In this section, we'll make a few simplifying assumptions, most importantly that the error on the sum and difference signals are uncorrelated,[15] and that the SNR is the same on both input channels.

To begin, we define noise versions of the sum and difference signals as

$$x_s(t) = s(t) + n_s(t) \qquad (7.109)$$
$$x_d(t) = d(t) + n_d(t) \qquad (7.110)$$

where the noise signals $n_s(t)$ and $n_d(t)$ are distributed with σ^2. The ratio is, thus, given by the equation

$$r(t) = \frac{x_d(t)}{x_s(t)} = \frac{d(t) + n_d(t)}{s(t) + n_s(t)} \qquad (7.111)$$

We define the true ratio $r_0(t) = d(t)/s(t)$ and the error signal as

$$n_r(t) = r(t) - r_0(t) = \frac{n_d(t) - r_0(t)n_s(t)}{s(t) + n_s(t)} \qquad (7.112)$$

This is a very complex noise term that depends on both of the component signals, not just their noise powers, or their ratio. Analysis is difficult, but some approximations can be made. In particular, if $\xi = |s(t)|^2 / \sigma^2$ is much greater than 1 (the noise is much weaker than the signal), then the error can be approximated with

$$n_r(t) = \frac{n_d(t)}{s(t)} - r_0(t) \frac{n_s(t)}{s(t)} \qquad (7.113)$$

and the variance of the noise term is approximated as

$$\sigma_r^2 = \frac{1}{\xi}\left(1 + r_0^2(t)\right) \qquad (7.114)$$

Next, we apply (7.108) to convert from ratio error variance to angle estimator error variance

$$\sigma_\theta^2 = \frac{1}{k_m^2 \xi}\left(1 + \left(k_m \frac{\theta}{\theta_{\text{bw}}}\right)^2\right) \qquad (7.115)$$

$$\sigma_\theta^2 = \theta_{\text{bw}}^2 \sigma_{\tilde{\theta}}^2 \qquad (7.116)$$

[15] One can see that this is the case if we assume that the sum and difference operators introduce no noise and that the noise on the two input signals is independent. The proof of this is left to the reader as an exercise.

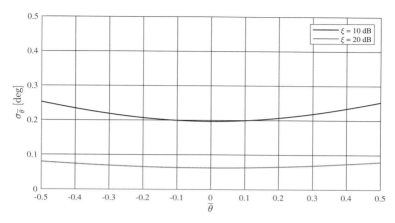

Figure 7.16 Direction-finding RMSE for a monopulse AOA receiver, as a function of $\bar{\theta}$.

This is plotted in Figure 7.16 as a function of $\bar{\theta}$, for $\xi = 10$ and 20 dB. Note that the error is plotted in units of normalized angle. From Figure 7.16, observe that if $\xi \geq 10$ dB, then the angular error is bounded by $\sigma_\theta \leq 0.25\theta_{bw}$, and if $\xi \geq 20$ dB the error drops to $\sigma_\theta \leq 0.08\theta_{bw}$. This demonstrates the ability of a monopulse receiver to significantly improve the angular accuracy of a directional antenna over a simple amplitude measurement.

7.7 PROBLEM SET

7.1 Use the provided MATLAB® functions to generate a test sequence for an Adcock array with spacing $d/\lambda = .5$, and $N = 100$ angular samples between $-90°$ and $90°$. For the test signals, use a simple sinusoid with $f = 1$ GHz, $t_s = 1/2f$, and $M = 100$ temporal samples at each position, a random starting phase (at each angular sample position), and a true angle of $5°$. Plot the amplitude response as a function of scan angle and time.

7.2 Add Gaussian noise with $\sigma^2 = .1$ to the test signal in the previous problem. Estimate the AOA.

7.3 Plot the CRLB of an Adcock antenna with $d/\lambda = .25$ and for a reflector antenna with $D/\lambda = 10$, as a function of ψ for $M\xi = 10$ dB.

7.4 Generate reference and test signals for a Watson-Watt receiver given a BPSK signal at $f = 300$ MHz, with a bandwidth of $B_t = 1$ MHz and an angle of $\theta = 15°$. Sample the signal at a rate of $t_s = 1/10B_t$ (five times the Nyquist sampling rate) for $M = 100$ samples. Add Gaussian noise with a variance of $\sigma^2 = .1$. Use the function aoa.watson_watt_df to estimate θ.

7.5 What is the CRLB of the Watson-Watt receiver in the previous problem?

7.6 Prove that (7.96) results from (7.95) and (7.94).

7.7 Consider a signal with center frequency $f_0 = 2$ GHz, bandwidth $B = 100$ kHz, and pulse length $t_p = 1$ ms, sampled at $f_s = 2f_0$ by a Doppler DF system with a rotation rate of $f_R = 1$ Hz and rotation radius $R = \lambda$. Assume that amplitude difference between the reference and test channels is negligible. Plot the CRLB as a function of ξ. What is the minimum SNR for which the angle accuracy is $\sigma_\psi \leq 10$ mrad. Repeat for rotation rates $f_R = 10$ and 100 Hz.

7.8 Plot the amplitude response for a two-element interferometer with $d = 3\lambda$ and $d = 4\lambda$. What is the maximum angle ψ for each system to avoid ambiguities? What is the maximum unambiguous angle if they are processed together to resolve ambiguities?

7.9 Consider a two-element interferometer with $d = \lambda/2$, and a source at $\psi = \pi/4$. The source amplitude is 1, and noise power is $\sigma^2 = .25$. Construct a baseband test signal consisting of a chirp (linear frequency modulated sweep) that starts at $f_0 = 0$ MHz and ends at $f_1 = 10$ MHz, sampled at $f_s = 100 MHz$ for $M = 100$ samples. Generate the test signal and random noise; use the provided code to estimate the AOA. Repeat for 1,000 random trials and compute the root mean squared error. Compare to the CRLB.

References

[1] D. Adamy, *EW 103: Tactical battlefield communications electronic warfare.* Norwood, MA: Artech House, 2008.

[2] D. Adamy, *EW 104: EW Against a New Generation of Threats.* Norwood, MA: Artech House, 2015.

[3] R. Poisel, *Electronic warfare target location methods.* Norwood, MA: Artech House, 2012.

References

[4] R. A. Poisel, *Electronic warfare receivers and receiving systems*. Norwood, MA: Artech House, 2015.

[5] W. Read, "Review of conventional tactical radio direction finding systems," DEFENCE RESEARCH ESTABLISHMENT OTTAWA (ONTARIO), Tech. Rep., 1989.

[6] H. L. Van Trees, *Optimum Array Processing: Part IV of Detection, Estimation and Modulation Theory*. Hoboken, NJ: Wiley-Interscience, 2002.

[7] N. O'Donoughue and J. M. F. Moura, "On the product of independent complex gaussians," *IEEE Transactions on Signal Processing*, vol. 60, no. 3, pp. 1050–1063, March 2012.

[8] S. V. Schell and W. A. Gardner, "High-resolution direction finding," *Handbook of statistics*, vol. 10, pp. 755–817, 1993.

[9] P. Q. C. Ly, "Fase and unambiguous direction finding for digital radar intercept receivers," Ph.D. dissertation, University of Adelaide, 2013.

[10] S. M. Sherman and D. K. Barton, *Monopulse principles and techniques*. Norwood, MA: Artech House, 2011.

[11] P. Q. C. Ly, "Fast and unambiguous direction finding for digital radar intercept receivers." Ph.D. dissertation, The University of Adelaide, 2013.

[12] F. Guo, Y. Fan, Y. Zhou, C. Xhou, and Q. Li, *Space electronic reconnaissance: localization theories and methods*. Hoboken, NJ: John Wiley & Sons, 2014.

[13] M. Pourhomayoun and M. Fowler, "Cramer-rao lower bounds for estimation of phase in LBI based localization systems," in *2012 Conference Record of the Forty Sixth Asilomar Conference on Signals, Systems and Computers (ASILOMAR)*, Nov 2012, pp. 909–911.

Chapter 8

Array-Based AOA

Chapter 7 introduces a number of specialized receiver architectures designed to estimate the direction of a received signal. This chapter focuses on a more general architecture. Given an array of spatially arranged antennas, each sampled and digitized for processing, one can employ a number of numerical techniques to estimate the direction of arrival of a signal.

In this chapter, we will discuss several classical approaches, including a standard beam former, an inverse filter (or *minimum-norm* filter), and a subspace-filtering approach referred to as multiple signal classifier (MUSIC). The field of available algorithms is rich and still expanding. It is our hope that these three results provide a firm basis with which to engage the literature.

We begin with a formulation of the general array problem, introduce the three solutions, and discuss the performance and limitations of each.

Much of the linear algebra and array processing formulation in this chapter is based on [1] and Chapter 9 of [2]. Interested readers are referred there for a more thorough discussion of array processing theory. Additional sources on each of the approaches, and variations thereof, are provided throughout this chapter.

8.1 BACKGROUND

We begin with a description of a linear array, as shown in Figure 8.1. This can be generalized to two- and even three-dimensional arbitrary arrays. For simplicity, we will contain ourselves to the one-dimensional case.

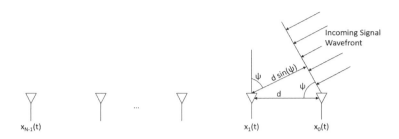

Figure 8.1 Linear array formulation, including broadside convention for AOA.

The array is constructed with a series of antenna elements that are processed coherently. Early generations of arrays used analog hardware to conduct the phase-shifting and combining operations, referred to as *analog beamforming*. However, as A/Ds have improved over time, the concept of digitizing each element of an array natively (that is to say, with direct I & Q sampling at the RF frequency) is becoming increasingly common, and all of the filtering and processing steps can often be done digitally, referred to as *digital beamforming* or, more generally, *array processing*, as shown in Figure 8.2. That being said, it is still common, particularly for large active arrays or for higher operating frequencies, to use a subarray architecture, in which individual sections of the array are pointed in a general direction using analog beamforming, and then the outputs are digitized and combined digitally. In this chapter, we are focused on estimation of the AOA of unknown signals, which is slightly different from the standard beamforming assumption where the angle is known (or commanded).

Readers interested in a more thorough discussion of phased arrays are directed to [1] for theoretical discussions focused on optimum processing of array signals, [3] for a detailed text on the process of designing a phased array system, and Chapter 13 of [4] for discussion of the use of phased arrays in radar, including operational examples and error tolerances in calibration.

8.1.1 Nonstandard Array Configurations

An array is loosely defined as any arbitrary set of antennas whose output is processed to provide directivity. Throughout this chapter, we focus on a specific instance of arrays called a *regular* array, which refers to arrays where the spacing between elements is uniform. These arrays lend well to analysis, and provide favorable sidelobe structures, but they require a very large number of antenna

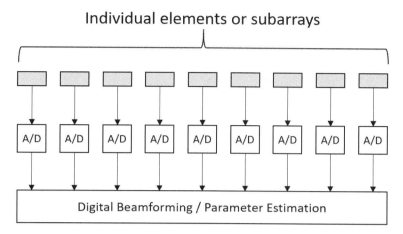

Figure 8.2 Digital array architecture.

elements, which can be prohibitive. Furthermore, this limits the use of large arrays on smaller platforms, such as UAVs, since they need to be mounted in a contiguous space. Relaxation of this uniform spacing allows for the array to have cutouts, such as where landing gear must be placed on a small UAV. For these reasons, and many others, there is a rich field of research into *irregular* arrays.

Some early examples include thinned arrays [5, 6], wherein a regular array is designed, and then elements are removed from it. Typically the antenna elements still exist, in order to avoid changes to the electromagnetic coupling that occurs between antennas, but the unused elements are connected to a ground plane instead of to a receiver. A similar approach is the use of a *random* array [7–9], with the difference being that the elements no longer sit on a grid, and the missing antenna elements are not populated in the array.

In addition to random or systematic approaches, genetic algorithms have also been applied to the problem of generating irregular arrays that seek to optimize some measure of goodness, such as peak or average sidelobe levels, or mainlobe width [10, 11].

More recently, renewed interest has been placed on sparse arrays [12, 13], which are similar in structure to random and thinned arrays, but are typically associated with either compressive sampling-based receivers [14, 15] or multiple-input multiple-output (MIMO) radars and communications systems [16–18].

8.2 FORMULATION

We begin with the time delay τ of a signal received from a plane wave arriving at θ degrees (ψ radians) at the kth antenna position. In the case of a uniform, linear array, the spacing between elements is constant, and is given by the variable d, and the time delay of the signal at the kth position is[1]

$$\tau_k = k\frac{d}{c}\sin(\psi) \tag{8.1}$$

and the corresponding phase shift (assuming a narrowband signal) is

$$\phi_k = 2\pi f_c \tau_k = \frac{2\pi d k}{\lambda}\sin(\psi) \tag{8.2}$$

where f_c is the signal's carrier frequency, and λ is the signal's wavelength. From this, we can define the kth input signal as

$$x_k[m] = s[m]e^{j\frac{2\pi d}{\lambda}k\sin(\psi)} \tag{8.3}$$

We collect these phase shifts into a *steering vector* defined

$$\mathbf{v}(\psi) = \left[1,\; e^{j\frac{2\pi d}{\lambda}\sin(\psi)},\, \ldots,\; e^{j\frac{2\pi d(N-1)}{\lambda}\sin(\psi)}\right] \tag{8.4}$$

For simplicity, we introduce the concept of u-space, which leads to a simpler notation for the steering vector

$$u = \sin(\psi) \tag{8.5}$$

$$\mathbf{v}(u) = \left[1,\; e^{j\frac{2\pi d}{\lambda}u},\, \ldots,\; e^{j\frac{2\pi d}{\lambda}(N-1)u}\right] \tag{8.6}$$

The enclosed MATLAB® code provides a utility for generating a steering vector, called make_steering_vector. An example is provided as follows.

```
d_lam = .5;
N = 11;
v_fun = array.make_steering_vector(d_lam,N);
```

---

[1] Our convention is that $\psi = 0$ corresponds to a broadside angle. Others, including Van Trees [1], define $\psi = 0$ as end-fire and $\psi = \pi/2$ as broadside. This difference manifests in (8.1) as a $\cos(\psi)$ term where we have $\sin(\psi)$.

8.2.1 Multiple Plane Waves

If there is more than one plane wave impinging on the array, then the received signal is the superposition of the individual waves. We collect the D AOAs ψ_d, steering vectors $\mathbf{v}(\psi_d)$, and received signals $s_d[m]$ into matrix form to compactly express the received signal across the array (in the presence of white Gaussian noise)

$$\mathbf{x}[m] = \mathbf{V}(\boldsymbol{\psi})\mathbf{s}[m] + \mathbf{n}[m] \tag{8.7}$$
$$\mathbf{V}[\boldsymbol{\psi}] = [\mathbf{v}(\psi_0), \ldots, \mathbf{v}(\psi_{D-1})] \tag{8.8}$$
$$\mathbf{s}[m] = [s_0[m], \ldots, s_{D-1}[m]]^T \tag{8.9}$$
$$\mathbf{n}[m] \sim \mathcal{CN}\left(0, \sigma_n^2 \mathbf{I}_N\right) \tag{8.10}$$

8.2.2 Wideband Signals

The progression of signals across an array is actually defined by a time delay at each element. For narrowband signals, this is equivalent to a phase shift of the incoming signals as modeled above. However, particularly for EW receivers, it is increasingly necessary to operate over a wide bandwidth. The simplest way to handle this is to *channelize* the receiver, as discussed in Chapter 5, and treat each channel separately with its own narrowband beamformer. The only downside to this approach is the requirement for additional RF hardware on each channel, and its inability to effectively handle signals that span multiple channels. Alternatively, if the system is fully digital and can be sampled at RF (rather than after down-conversion), then beamforming can be done digitally, and channelization is achieved simply with additional computational power.

For the remainder of this chapter, we assume narrowband beamforming.

8.2.3 Array Beamforming

Given the steering vector defined above, it is straightforward to compute a beam-former that will focus on signals from a desired direction of arrival (ψ_0 or u_0) based on the steering vector for a signal from that direction [1, 19, 20].

Each of the input signals is multiplied by a complex weight h_k, and the output is summed together. This is written formally

$$y[m] = \mathbf{h}^H \mathbf{x}[m] = \sum_{n=0}^{N-1} x_k[m] h_k^* \qquad (8.11)$$

A straightforward approach to computing the beamformer weights is to define $\mathbf{h}$ as the steering vector from a signal at the desired angle ψ_0.

$$\mathbf{h}(\psi_0) = \mathbf{v}(\psi_0) \qquad (8.12)$$

Using this definition of the steering vector, the output of the beamformer (assuming a single input signal from direction ψ) is given as

$$y[m] = s[m] \mathbf{h}^H(\psi_0) \mathbf{v}(\psi) \qquad (8.13)$$

$$= s[m] \sum_{k=0}^{N-1} e^{j \frac{2\pi d}{\lambda} k [\sin(\psi) - \sin(\psi_0)]} \qquad (8.14)$$

This is simplified to

$$y[m] = s[m] \frac{\sin\left(\frac{N\pi d}{\lambda}[\sin(\psi) - \sin(\psi_0)]\right)}{\sin\left(\frac{\pi d}{\lambda}[\sin(\psi) - \sin(\psi_0)]\right)} \qquad (8.15)$$

The second term is often referred to as the *Array factor*, and defines the beampattern of an array with a given set of beamformer weights h_k. More formally, it is defined as

$$AF(\psi, \psi_0) = \frac{\mathbf{h}^H(\psi_0) \mathbf{v}(\psi)}{\|\mathbf{h}(\psi_0)\| \, \|\mathbf{v}(\psi)\|} \qquad (8.16)$$

where the magnitude terms in the denominator serve to normalize the output to a peak value of 1. For the case of a uniform linear array, and standard beamformer in (8.12), the array factor is given directly as:

$$AF(\psi, \psi_0) = \frac{\sin\left(\frac{N\pi d}{\lambda}(\sin(\psi) - \sin(\psi_0))\right)}{N \sin\left(\frac{\pi d}{\lambda}(\sin(\psi) - \sin(\psi_0))\right)} \qquad (8.17)$$

$$AF(u, u_0) = \frac{\sin\left(\frac{N\pi d}{\lambda}(u - u_0)\right)}{N \sin\left(\frac{\pi d}{\lambda}(u - u_0)\right)} \qquad (8.18)$$

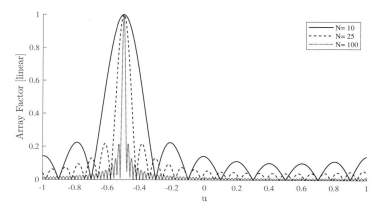

Figure 8.3 Array factor plot for a linear array with half-wavelength spacing ($d = \lambda/2$) and various array lengths (N elements). Increasing array length corresponds to narrower mainlobe and alters sidelobes.

We plot the array factor for a linear array with $N = 10$, 25, and 100 elements in Figure 8.3 as a function of the AOA ψ, for an array steered to $-30°$ ($\psi_0 = -\pi/6$, $u_0 = -.5$) and $\mathbf{h}(\psi_0) = \mathbf{v}(\psi_0)$.

This is generated in the accompanying MATLAB® code with a call to compute_array_factor_ula, which implements (8.17) directly.

```
psi_0 = -30*(pi/180);
psi = linspace(-pi/2,pi/2,101);
d_lam = .5;
N=25;
af_ula = array.compute_array_factor_ula(d_lam,N,psi,psi_0);
```

A more general form is useful for arbitrary beamformers or complex received signals is provided in the function array.compute_array_factor.

```
d_lam=.5;
N=25;
v = array.make_steering_vector(d_lam,N);
psi_0 = -30*(pi/180);
h = v(psi_0);
psi = linspace(-pi/2,pi/2,101);
af_arbitrary = array.compute_array_factor(v,h,psi);
```

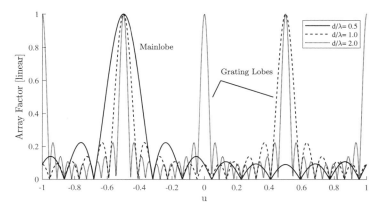

Figure 8.4 Array factor for a uniform linear array with $N = 25$ elements and various interelement spacing d. When $d > \lambda/2$ grating lobes are observed, which represents ambiguity in the array factor.

In general, as with the interferometer from Chapter 7, the interelement spacing should be no larger than $d = \lambda/2$. It can be relaxed slightly if the elements have a directional beampattern, or if signals beyond a certain azimuth angle can be ignored, but half-wavelength spacing is robust to signals from any incoming AOA. To illustrate this, we plot the beampattern for an array with $N = 10$ elements steered to $u_0 = -0.5$ and various interelement spacings between $d = \lambda/2$ and $d = 2\lambda$ in Figure 8.4; note the existence, and location, of grating lobes for all cases where $d > \lambda/2$.

8.2.4 Nonisotropic Element Patterns

The array factor in (8.16) implicitly assumes that the elements that make up the array all have a perfectly omnidirectional response. In reality, this is rarely the case. If we define the element response of the kth antenna $a_k(\psi)$ and collect that into an array $\mathbf{a}(\psi)$, then the array factor is written as

$$AF(\psi, \psi_0) = \mathbf{h}^H(\psi_0)\left[\mathbf{a}(\psi) \odot \mathbf{v}(\psi)\right] \qquad (8.19)$$

where $\odot$ is the element-wise product, called the *Hadamard product*. If all antennas have the same pattern, then $a_k(\psi) = a(\psi)$ and the array factor simplifies to

$$AF(\psi, \psi_0) = a(\psi)\mathbf{h}^H(\psi_0)\mathbf{v}(\psi) \qquad (8.20)$$

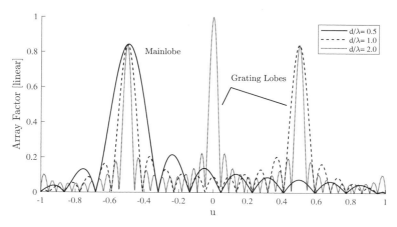

Figure 8.5 Replot of Figure 8.4, with an element pattern defined by $\cos^{1.2}(\psi)$.

Figure 8.5 plots a beampattern with grating lobes, as in figure 8.4, but with the addition of an element pattern defined by $\cos^{1.2}(\psi)$. This is implemented with an optional argument to `compute_array_factor_ula`:

```
el_pat_fun = @(psi) cos(psi).^1.2;
af = array.compute_array_factor_ula(d_lam,N,psi,psi_0,el_pat_fun);
```

Note that the element pattern did not appreciably alter the shape or location of any of the dominant lobes in Figure 8.5 (with the exception of slightly shifting the peak of the mainlobe for the $d/\lambda = 0.5$ case); the dominant effect is that the lobes far from broadside are reduced. For the $d/\lambda = 2$ case, the grating lobe located at $u = 0$ now appears to be the dominant return, and could result in an erroneous position estimate. For this reason, it is important to be careful about estimation when grating lobes are a possibility, particularly if there is a nonuniform element pattern.

8.2.5 Gain and Beamwidth

Gain is defined as an increase in SNR. In the context of array processing, it is computed when the array is properly steered ($\psi = \psi_0$), and is equivalent to the number of antenna elements in the array N (in linear space). To prove this, we consider a noisy signal

$$\mathbf{x}[m] = s[m]\mathbf{v}(\psi) + \mathbf{n}[m] \qquad (8.21)$$

where $\mathbf{n}[m]$ is a zero-mean complex Gaussian with covariance given by $\sigma_n^2 \mathbf{I}_N$ and the SNR on each input antenna is defined $\xi = |s[m]|^2 / \sigma_n^2$. After beamforming, the output is

$$y[m] = s[m]\mathbf{h}^H(\psi_0)\mathbf{v}(\psi) + \mathbf{h}^H(\psi_0)\mathbf{n}[m] \qquad (8.22)$$

We assume $\psi_0 = \psi$ (the array is properly steered) and compute the SNR

$$\xi_{\text{out}} = \frac{|s[m]\mathbf{h}^H(\psi)\mathbf{v}(\psi)|^2}{E\left\{|\mathbf{h}^H(\psi)\mathbf{n}[m]|^2\right\}} \qquad (8.23)$$

$$= \frac{|s[m]|^2 |\mathbf{v}^H(\psi)\mathbf{v}(\psi)|^2}{\mathbf{h}^H(\psi) E\{\mathbf{n}[m]\mathbf{n}^H[m]\} \mathbf{h}(\psi)} \qquad (8.24)$$

$$= \frac{N^2 |s[m]|^2}{N \sigma_n^2} = N \xi_{\text{in}} \qquad (8.25)$$

Thus, the SNR after beamforming (ξ_{out}) is N times the input SNR at each element (ξ_{in}). The half-power beamwidth, which we denote δ_u (for u-space) is the width of the mainlobe, as measured by the points where the received power is half of the peak level (3 dB below the peak in log space). It is a useful measure of resolution, as two equally powered signals separated by $|u_2 - u_1| < \delta_u$ are not distinguishable.

For a linear array with half-wavelength spacing, this occurs at

$$\delta_u = .89 \frac{2}{N-1} \qquad (8.26)$$

More generally, given spacing d and wavelength λ, the beamwidth is

$$\delta_u = .89 \frac{\lambda}{d(N-1)} \qquad (8.27)$$

Figure 8.6 shows the array factor for a scene with two sources placed on equal sides of $u = 0$, with spacing δ_u. At the point of intersection, each signal is at -3 dB relative to the peak return (half power).

8.2.6 Array Tapers

Thus far, we have assumed that the beamformer $\mathbf{h}(\psi_0)$ has uniform amplitude across all of its elements. If this assumption is relaxed, then it is possible to apply

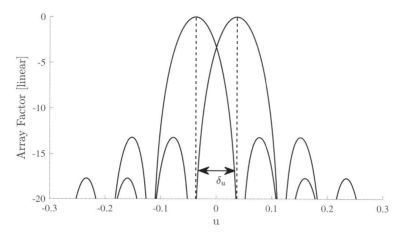

Figure 8.6 Array factor for a standard beamformer responding to two signals separated by the array beamwidth.

one of the many well-studied tapers to control the location of sidelobes, or to generate nulls.

This section is not exhaustive, and merely presents a small number of tapers designed to suppress the first few sidelobes, and based on weighted sums of cosine terms. More general tapers are discussed in Chapter 3 of [1]. Of note are the Taylor [21] and Bayliss [22] tapers used to control sidelobes for sum and difference beams in monopulse processing.

In this section, we expand the definition of the standard beamformer $\mathbf{h}$ to include nonuniform weights w_k collected into a vector $\mathbf{w}$.

$$\mathbf{h}(\psi_0) = \mathbf{w} \odot \mathbf{v}(\psi_0) \tag{8.28}$$

Figure 8.7 plots the value of w_k for four separate tapers, a uniform, cosine, cosine-squared (or Hann), and a Hamming, as well as the resultant array factors for an $N = 11$ element linear array, computed via a Fourier transform of the weights. Each of these tapers is described individually in the following text.

The *cosine window* exhibits a complementary structure to a uniform (unweighted) taper, with its nulls corresponding to locations of the uniform taper's peaks, and vice versa. The mainlobe is broadened so that the first null falls where the first peak of the untapered pattern occurs; this results in reduced sidelobe levels,

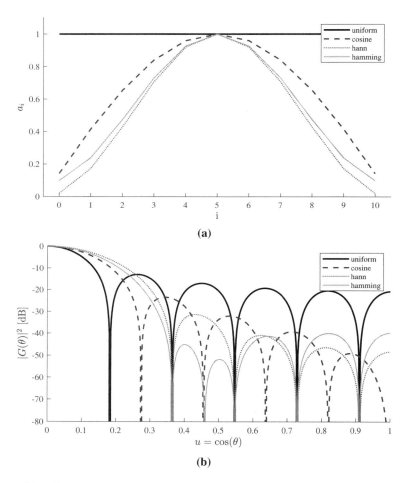

Figure 8.7 Effects of some tapered beamformers: (a) taper weights, (b) corresponding beampatterns.

Array-Based AOA

Table 8.1
Common Amplitude Tapers

Taper	HPBW (1)	SLL
Uniform	$0.89 \frac{2}{N-1}$	-13.0 dB
Cosine	$1.18 \frac{2}{N-1}$	-23.5 dB
Hann	$1.44 \frac{2}{N-1}$	-31.4 dB
Hamming	$1.31 \frac{2}{N-1}$	-39.5 dB

(1) Half-power beamwidth (HPBW) expressed in units of $u - u_0$.

with the first sidelobe occurring as -23.5 dB, rather than -13.0 dB. It is defined as

$$w_k = \sin\left(\frac{\pi}{2N}\right) \cos\left(\frac{\pi}{N}\left(k - \frac{N-1}{2}\right)\right) \quad k = 0, \ldots, N-1 \quad (8.29)$$

The *Hann window*,[2] also referred to as a *raised cosine*, is designed to suppress the first sidelobe of its response, and indeed the first sidelobe occurs at -31.4 dB, but the mainlobe is broadened to include the entire first sidelobe of the untapered array. It is defined as

$$w_k = \cos^2\left(\frac{\pi}{N}\left(k - \frac{N-1}{2}\right)\right) \quad k = 0, \ldots, N-1 \quad (8.30)$$

The *Hamming window* has coefficients designed to suppress the first sidelobe of the former's response. This results in a peak of -39.5 for largest sidelobe of the Hamming window. It is defined as

$$w_k = 0.54 - 0.46 \cos\left(\frac{2\pi k}{N-1}\right) \quad k = 0, \ldots, N-1 \quad (8.31)$$

Table 8.1 summarizes the performance of the tapers in this section to illustrate the differences in mainlobe width and sidelobe amplitude.

8.2.7 Two-Dimensional Arrays

To extend the results in this chapter for two-dimensional direction of arrival, we first define the three-dimensional coordinate system, which will rely on a pair of

---

[2] Hann windows are sometimes erroneously referred to as Hanning, likely due to confusion with Hamming windows.

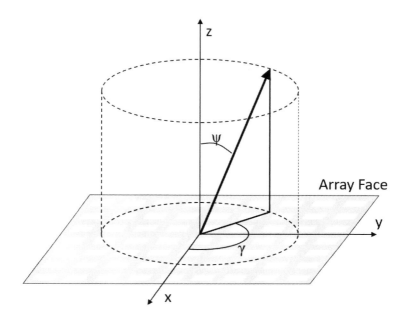

Figure 8.8 Coordinate system for planar arrays, describing the cone angle ψ and rotation angle γ in relation to the planar array's face.

angles, the cone angle and the rotation angle. The cone angle, ψ, is the angle between the incoming signal's propagation vector and the array's broadside (the vector normal to the array face for a planar array), and the rotation angle, γ, which is defined as the angle between the $+x$ axis of the array face, and the projection of the incoming signal's propagation vector onto the array face.[3] This is shown graphically in Figure 8.8.

$$u_x = \sin(\psi)\cos(\gamma) \qquad (8.32)$$
$$u_y = \sin(\psi)\sin(\gamma) \qquad (8.33)$$

All of the same logic and results from the beam pattern and taper discussion in this section can be extended to the case of a 2-D planar array. If the array geometry

---

[3] There are many conventions for 3-D angular geometry. θ and ϕ are commonly used, but we use ψ and γ to avoid confusion with the fact that elsewhere in this text, ϕ is used for complex phase.

is separable (such as a regular grid), and the tapers are separable ($w_{i,j} = w_{x,i} w_{y,j}$), then analysis can be conducted independently in the two orthogonal dimensions (u_x and u_y).

8.3 SOLUTION

Consider the received signal vector $\mathbf{x}[m]$ for $m = 0, \ldots, M-1$ temporal snapshots and N antenna elements. For generality, we assume that D plane waves are impinging on the array. The snapshot vector is given as

$$\mathbf{x}[m] = \mathbf{V}(\boldsymbol{\psi})\mathbf{s}[m] + \mathbf{n}[m] \tag{8.34}$$

Going forward, we will use $\mathbf{V}$ as shorthand for $\mathbf{V}(\boldsymbol{\psi})$. In this chapter, we assume that the noise vector $\mathbf{n}[m]$ is an IID complex Gaussian random vector with zero mean and covariance matrix $\sigma_n^2 \mathbf{I}_N$.

An important related problem is the estimation of the number of signals impinging upon an array. For simplicity, we assume that D is known. The problem of estimating D, and its importance to AOA estimation algorithms, has been studied for more than 30 years [23, 24].

8.3.1 Signal Models

As with any estimation problem, the choice of signal model is crucial to the derivation of an optimal estimator, and to analysis of errors. In the case of direction finding, the source signal $s[m]$ is often unknown, but whether this unknown value is taken to be a random process, or a deterministic instantiation of a random process will vary the results. In general, the former result is termed *unconditioned* estimation, in that no conditions are placed on the distribution of $s[m]$. The latter is referred to as *conditioned* in that all of the probability distributions are conditioned upon the signal $s[m]$ that was transmitted.

As we will see, the end result (in terms of CRLB performance bounds) is the same for both data models, but the ML estimator is unique. We will deal briefly with the *unconditioned* case, but will spend more time deriving the *conditioned* ML estimate. The end result, in either case, is an optimization problem without a closed-form solution. Thus, after the ML Estimate is derived, we present several foundational approaches to achieving an AOA estimate: beamscanning, inverse filtering, and subspace-based methods. This chapter is not exhaustive, and is merely presented as a starting point from which the reader can delve into more complex solutions.

8.3.2 Maximum Likelihood Estimation

The maximum likelihood solution to estimating the direction of arrival for all D sources depends on the assumptions made about $\mathbf{s}[m]$. If we assume that $\mathbf{s}[m]$ is a random sample with some covariance matrix $\mathbf{C_s}$, then arrive at an ML estimator called the *stochastic* or *unconditioned* ML estimate of $\boldsymbol{\psi}$. We will focus our attention on the latter, as its signal model is more similar to that taken in prior chapters. First, however, we will briefly summarize the motivation and solution for the *unconditioned* ML estimator.

For an *unconditioned* ML estimator, we model the signal vector $\mathbf{s}[m]$ as a random vector with zero mean and covariance matrix $\mathbf{C_s}$. From this, we arrive at the signal model

$$\mathbf{x}[m] \sim \mathcal{CN}\left(\mathbf{0}_{N,1}, \mathbf{V}\mathbf{C_s}\mathbf{V}^H + \sigma_n^2 \mathbf{I}_N\right) \tag{8.35}$$

The approach to derive the ML estimator follows the general guideline in Chapter 6, with a slight modification. Both the signal covariance matrix $\mathbf{C_s}$ and directions of arrival $\boldsymbol{\psi}$ are unknown. The solution, according to [1], is to first estimate the covariance matrix of $\mathbf{s}[m]$ functionally,

$$\widehat{\mathbf{C}}_\mathbf{s} \triangleq \mathbf{V}^\dagger \left[\mathbf{R_x} - \sigma_n^2 \mathbf{I}_N\right] \left[\mathbf{V}^\dagger\right]^H \tag{8.36}$$

and insert this solution into the log likelihood function

$$\widehat{\boldsymbol{\psi}} = \arg\max_{\boldsymbol{\psi}} \; -\ln\left|\mathbf{P_V}\mathbf{R_x}\mathbf{P_V}^H + \sigma_n^2 \mathbf{P_V^\perp}\right| - \frac{1}{\sigma_n^2}\mathrm{Tr}\left\{\mathbf{P_V^\perp}\mathbf{R_x}\right\} \tag{8.37}$$

where $\mathbf{R_x}$ is defined as the sample covariance of $\mathbf{x}[m]$

$$\mathbf{R_x} = \frac{1}{M}\sum_{m=0}^{M-1} \mathbf{x}[m]\mathbf{x}^H[m] \tag{8.38}$$

$\mathbf{P_V}$ is the projection operator onto the subspace $\mathbf{V}$

$$\mathbf{P_V} = \mathbf{V}\left(\mathbf{V}^H\mathbf{V}\right)^{-1}\mathbf{V}^H \tag{8.39}$$

and $\mathbf{P_V^\perp}$ is the orthogonal projection. We will not dwell on this estimator very much, except to say that it does exist, and that is it more accurate than the *conditional* estimator we will derive next when (a) the signals to be estimated are closer than the null-to-null beamwidth of the array, (b) the sources are correlated, and (c) the signals have widely divergent strengths (some are much weaker than others). If any of these are the case, then an *unconditioned* ML estimator based on (8.37) is more likely to be accurate. Interested readers are referred to Section 8.5 of [1] for a detailed derivation, based on [25].

The signal model for *conditional* ML estimation is to consider s[m] to be a complex-valued unknown, but deterministic, parameter. In this case, the data snapshot distribution is

$$\mathbf{x}[m] \sim \mathcal{CN}\left(\mathbf{V}\mathbf{s}[m], \sigma_n^2 \mathbf{I}_N\right) \tag{8.40}$$

and the log likelihood function is

$$\ell\left(\mathbf{x}|\boldsymbol{\psi}, \mathbf{s}\right) = -MN\ln\sigma_n^2 - \frac{1}{\sigma_n^2} \sum_{m=0}^{M-1} |\mathbf{x}[m]\mathbf{V}\mathbf{s}[m]|^2 \tag{8.41}$$

The first term does not vary with either of the unknown parameters, so it can safely be ignored. Thus, we can equivalently minimize

$$\ell_1\left(\mathbf{x}|\boldsymbol{\psi}, \mathbf{s}\right) = \sum_{m=0}^{M-1} |\mathbf{x}[m] - \mathbf{V}\mathbf{s}[m]|^2 \tag{8.42}$$

To solve this, we take a similar approach to the *unconditional* case, and first fix $\boldsymbol{\psi}$. The ML estimate of s[m]

$$\widehat{\mathbf{s}}[m] \triangleq \left[\mathbf{V}^H\mathbf{V}\right]^{-1}\mathbf{V}^H\mathbf{x}[m] \tag{8.43}$$

We insert this solution into (8.42), and simplify to

$$\begin{aligned}\ell_1\left(\mathbf{x}|\boldsymbol{\psi}\right) &= \sum_{m=0}^{M-1}\left|\mathbf{x}[m] - \mathbf{V}\left[\mathbf{V}^H\mathbf{V}\right]^{-1}\mathbf{V}^H\mathbf{x}[m]\right|^2 & (8.44)\\ &= \sum_{m=0}^{M-1}|\mathbf{x}[m] - \mathbf{P_V}\mathbf{x}[m]|^2 = \sum_{m=0}^{M-1}\left|\mathbf{P_V^\perp}\mathbf{x}[m]\right|^2 & (8.45)\end{aligned}$$

Next, we use the matrix identity $\mathbf{a}^H\mathbf{a} = \text{Tr}\{\mathbf{a}\mathbf{a}^H\}$ to further simplify to

$$\ell_1(\mathbf{x}|\boldsymbol{\psi}) = \sum_{m=0}^{M-1} |\mathbf{P}_\mathbf{V}^\perp \mathbf{x}[m]|^2 = \sum_{m=0}^{M-1} \left(\mathbf{P}_\mathbf{V}^\perp \mathbf{x}[m]\right)^H \left(\mathbf{P}_\mathbf{V}^\perp \mathbf{x}[m]\right) \quad (8.46)$$

$$= \sum_{m=0}^{M-1} \text{Tr}\left\{\mathbf{P}_\mathbf{V}^\perp \mathbf{x}[m]\mathbf{x}^H[m]\mathbf{P}_\mathbf{V}^\perp\right\} \quad (8.47)$$

$$= \text{Tr}\left\{K\mathbf{P}_\mathbf{V}^\perp \left(\frac{1}{K}\sum_{m=0}^{M-1} \mathbf{x}[m]\mathbf{x}^H[m]\right) \mathbf{P}_\mathbf{V}^\perp\right\} \quad (8.48)$$

$$= K\text{Tr}\left\{\mathbf{P}_\mathbf{V}^\perp \mathbf{R}_\mathbf{x} \mathbf{P}_\mathbf{V}^\perp\right\} = K\text{Tr}\left\{\mathbf{P}_\mathbf{V}^\perp \mathbf{R}_\mathbf{x}\right\} \quad (8.49)$$

where the simplification on the final line relies on the fact that both $\mathbf{R}_\mathbf{x}$ and $\mathbf{P}_\mathbf{V}^\perp$ are Hermitian (that is, they are equal to their conjugate transposes) to switch their order, and then the fact that the projection operator is idempotent ($\mathbf{PP} = \mathbf{P}$). From this simplification, we can see that the maximum likelihood estimate $\boldsymbol{\psi}$ can be found with the minimization of the projection of the sample covariance matrix $\mathbf{R}_\mathbf{x}$ onto the orthogonal subspace of $\mathbf{V}$

$$\widehat{\boldsymbol{\psi}} = \arg\min_{\boldsymbol{\psi}} \text{Tr}\left\{\mathbf{P}_\mathbf{V}^\perp \mathbf{R}_\mathbf{x}\right\} \quad (8.50)$$

or, to equivalently maximize the projection onto the subspace of $\mathbf{V}$

$$\widehat{\boldsymbol{\psi}} = \arg\max_{\boldsymbol{\psi}} \text{Tr}\left\{\mathbf{P}_\mathbf{V} \mathbf{R}_\mathbf{x}\right\} \quad (8.51)$$

Equivalently, we can express the optimization via the eigenvalue decomposition of $\mathbf{R}_\mathbf{x}$

$$\mathbf{R}_\mathbf{x} = \sum_{k=0}^{N-1} \lambda_k \mathbf{u}_k \mathbf{u}_k^H \quad (8.52)$$

$$\widehat{\boldsymbol{\psi}} = \arg\max_{\boldsymbol{\psi}} \sum_{k=0}^{N-1} \lambda_k |\mathbf{P}_\mathbf{V} \mathbf{u}_k|^2 \quad (8.53)$$

In plain language, this estimator takes a proposed set of AOAs, $\boldsymbol{\psi}$, projects each of the eigenvectors onto the subspace defined by those directions of arrival, and weights the sum of those projections by the associated eigenvalues.

A closed-form solution for (8.51) does not exist, but there are many algorithms proposed to arrive at an estimate based on that optimization, including nonlinear programming and least squares approximation approaches to arrive at an estimate of $\widehat{\psi}$. Interested readers are referred to Section 3.2 of [26], and Section 8.6 of [1]. More generally, Boyd has written a comprehensive resource on numerical optimization techniques [27].

8.3.3 Beamformer Scanning

Perhaps the most intuitive approach to AOA estimation is to take a beamformer $\mathbf{h}(\psi_0)$, scan it across a set of candidate angles ψ, and pick the D highest peaks as the estimates $\widehat{\psi}_1, \ldots, \widehat{\psi}_D$. The power at each angle is computed as

$$\widehat{P}(\psi) = \frac{1}{M} \sum_{m=0}^{M-1} \left| \mathbf{h}^H(\psi) \mathbf{x}[m] \right|^2 \tag{8.54}$$

With some manipulation, we can show that this is rewritten as a simple projection on the sample covariance matrix $\mathbf{R_x}$

$$\widehat{P}(\psi) = \mathbf{h}^H(\psi) \mathbf{R_x} \mathbf{v}(\psi) \tag{8.55}$$

This is sometimes called a *beamscan* algorithm when $\mathbf{h}(\psi)$ is taken to be the steering vector $\mathbf{v}(\psi)$. The principle drawback of this approach is that the angular resolution is dictated by array length (and, when a taper is used, the beamwidth broadening caused by that taper).

This is computed very simply with the following MATLAB® code, which assumes the data matrix x is organized into N rows (one for each antenna element) and M columns (one for each snapshot). We first generate the steering vector, and then the test signal (for simplicity, we omit all of the steps taken to generate a noisy test signal), construct the matrix V of steering vectors for each of the scan positions ψ, and finally evaluate (8.54).

```
% Generate steering vector
d_lam = .5;
N = 25;
v = array.make_steering_vector(d_lam,N);

% Generate steering vector matrix
psi_vec = linspace(psi_min,psi_max,N_test_pts);
```

```
V = v(psi_vec)/sqrt(N);

% Compute the beamscan image at each scan angle
P = sum(abs(x'*V).^2,1)/M;
```

where x is the $N \times M$ received data vector for N antenna positions and M time samples. Alternative beamformers can be found in the adaptive beamforming literature. Among them is the *minimum variance distortionless response* (MVDR) beamformer, also referred to as a Capon beamformer [28], which we discuss here.

In adaptive filtering literature, the MVDR filter is an adaptive filter **h** that can whiten non-Gaussian noise, but avoids distortion of the desired signal (hence the name). In the context of AOA estimation, this filter will minimize additive interference between neighboring sources. For example, if two sources are close by, their signals will overlap. The MVDR beamformer, is written as

$$\mathbf{h}_{MVDR}(\psi) = \frac{\mathbf{R}_{\mathbf{x}}^{-1}\mathbf{v}(\psi)}{\mathbf{v}^H(\psi)\mathbf{R}_{\mathbf{x}}^{-1}\mathbf{v}(\psi)} \tag{8.56}$$

One can see from the denominator that it is normalized by the projection of the spectral matrix onto the current steering vector. If we insert this into (8.54), the result is

$$\widehat{P}_{MVDR}(\psi) = \frac{1}{M}\sum_{m=0}^{M-1}\left|\mathbf{h}_{MVDR}^H(\psi)\mathbf{x}[m]\right|^2 \tag{8.57}$$

$$= \frac{1}{M}\sum_{m=0}^{M-1}\left|\frac{\mathbf{v}^H(\psi)\mathbf{R}_{\mathbf{x}}^{-1}\mathbf{x}[m]}{\mathbf{v}^H(\psi)\mathbf{R}_{\mathbf{x}}^{-1}\mathbf{v}(\psi)}\right|^2 \tag{8.58}$$

$$= \frac{1}{\mathbf{v}^H(\psi)\mathbf{R}_{\mathbf{x}}^{-1}\mathbf{v}(\psi)} \tag{8.59}$$

The following MATLAB® code implements an MVDR beamformer. The first piece constructs the sample covariance matrix C and decomposes it,[4] and the second computes (8.59).

```
% Generate sample covariance matrix
[N,M] = size(x);
```

---

[4] Direct inversion of a covariance matrix is costly, but since the matrix is positive semidefinite, QR decomposition can reduce the complexity, particularly if the inversion occurs repeatedly. This is handled natively with the decomposition function in MATLAB.

```
C = zeros(N,N);
for idx_m = 1:M
    C = C + x(:,idx_m)*x(:,idx_m)'/M;
end
C_d = decomposition(C,'qr');

% MVDR Beamformer
P_mvdr = zeros(size(psi_vec));
for idx_psi = 1:numel(psi_vec)
    vv = v(psi_vec(idx_psi))/sqrt(N);
    P_mvdr(idx_psi) = 1./abs((vv'/C_d)*vv);
end
```

Because the MVDR is an inverse filter, it has the potential to present much narrower resolution than the standard beamformer.

The accompanying MATLAB® code provides functions for the beamscan and MVDR beamscan estimators, under the functions array.beamscan and array.beamscan_mvdr. They can be called with the following script, assuming x and v are defined as in the prior code snippets.

```
% Generate Beamscan Images
psi_max = pi/2;
N_pts = 1001;
[P,psi_vec] = array.beamscan(x,v,psi_max,N_pts);
[P_mvdr,psi_vec_mvdr] = array.beamscan_mvdr(x,v,psi_max,N_pts);
```

We plot the power levels for the beamformer output $\widehat{P}(\psi)$ and $\widehat{P}_{MVDR}(\psi)$ in Figure 8.9 for a case with two sources closely separated, with $N = 10$ elements in the array, and $\xi = 20$ dB. The improved resolution for an MVDR beamformer is immediately evident in the narrow spikes located at each of the source angles, while the standard beamformer is limited by the beamwidth of the array, and is unable to adequately resolve them. Another apparent result is that the standard beamformer has sidelobe levels that begin -13 dB below the peak; this can mask weaker sources. Although tapers are an effective tool for suppressing these sidelobes, they further broaden the mainlobe, limiting resolution. The MVDR, by contract, has returns around -25 dB for nonsource angles, which is a function of the SNR (in this case, 20 dB).

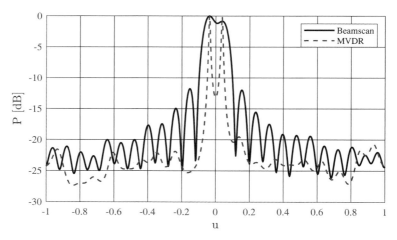

Figure 8.9 Power spectrum from a standard and MVDR beamscan approach for two closely spaced sources with $\xi = 10$ dB, showing the improved separability of MVDR (when ξ is large).

Example 8.1   VHF Push-to-Talk Radio DF

The provided MATLAB® code contains a sample data set from a set of VHF push-to-talk radios at a standoff range of approximately 50 km, the file can be found with the path examples/ex8_1.mat. The data was collected from an array with $N = 25$ elements and $d = .5\lambda$. Process the data to determine the angle of arrival for the D=3 source signals contained therein.

The first step is to load the data into memory

```
load examples/ex8_1.mat;
```

which loads the variables y (received data vector), D, N, M, and d_lam. The first step in DOA estimation is setting up the array steering vector and compute the beamscan image. First, we attempt the standard beamformer, with 1,001 sample points between $-\pi/2$ and $\pi/2$.

```
v = array.make_steering_vector(d_lam,N);
[P,psi_vec] = array.beamscan(v,y,pi/2,1001);
```

The next step is to find the peaks. If you have the *signal processing toolbox* installed, this is as simple as calling the command findpeaks, sorting the returned peaks, and grabbing the D first peaks.

```
[peak_vals,peak_idx] = findpeaks(P);
[~,sort_idx] = sort(peak_vals,'descend');
psi_soln = psi_vec(sort_idx(1:D));
```

For readability, we convert to degrees and print the result.

```
> th_soln = 180*psi_soln/pi
th_soln =
    -30.0600    20.5200    12.9600
```

Note that the first two peaks (at $-30°$ and $20°$) are picked out with moderate error (the second is off by $0.52°$), but the third is picking off a sidelobe of the combined lobe for the two sources near $20°$. Hence, it is likely a beamscan imager without knowledge of D would erroneously assume in this case that only two sources exist. Let us repeat the exercise with the MVDR beamformer.

```
[P_mvdr,psi_vec] = array.beamscan_mvdr(y,v,pi/2,1001);
[peak_val,peak_idx] = findpeaks(P_mvdr,psi_vec);
[~,sort_idx] = sort(peak_val,'descend');
psi_soln_mvdr = peak_idx(sort_idx(1:D));
```

```
> th_soln_mvdr = 180*psi_soln_mvdr/pi
th_soln_mvdr =
    19.9800    23.9400    -30.0600
```

Here, the results accurately line up with the three source locations. We plot the output of these two beamscan beamformers, and the selected peaks, in Figure 8.10.

8.3.3.1 Robust Capon

A well-known limitation of the MVDR beamformer is its sensitivity to mismatch between the true AOA ψ, and the nearest test point ψ_k, or from array calibration errors that cause differences between the true and assumed steering vectors for a given angle [29]. In order to alleviate this constraint, a family of *robust* beamformers, often referred to as *robust Capon* have been proposed. The simplest versions modify the sample covariance matrix with diagonal loading, but the literature is

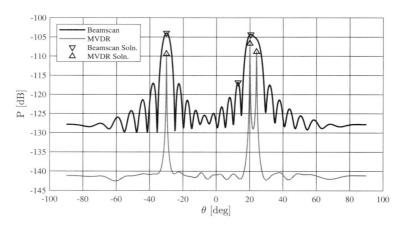

Figure 8.10 Beamscan and MVDR estimation of $D = 3$ source AOAs, from Example 8.1.

extensive and even includes recent examples that cannot be expressed in closed-form. For a good review of robust Capon beamformer literature, interested readers are referred to the introductory sections of [29].

8.3.4 Subspace-Based Methods

Recall from the discussion of maximum likelihood estimation that separation of the signal into a D-dimensional subspace containing the signals, and $N - D$-dimensional subspace containing noise alone (achieved by projecting the sample covariance matrix $\mathbf{R_x}$ onto the subspace spanned by $\mathbf{V}$). This section discusses methods to more fully take advantage of that decomposition. Recall the snapshot spectral matrix model

$$\mathbf{C_x} = \mathbf{V}\mathbf{C_f}\mathbf{V}^H + \sigma_n^2 \mathbf{I}_N \tag{8.60}$$

Assuming that the number of signals D is known, we can estimate the D source positions with a decomposition of $\mathbf{C_x}$. First, we note the eigenvalue representation

$$\mathbf{C_x} = \sum_{k=1}^{N} \lambda_k \mathbf{u}_k \mathbf{u}_k^H \tag{8.61}$$

This can be decomposed into source and noise subspaces (assuming that the eigenvalues are sorted in descending order)

$$\mathbf{C_x} = \underbrace{\sum_{k=0}^{D-1} \lambda_k \mathbf{u}_k \mathbf{u}_k^H}_{\text{source}} + \underbrace{\sum_{k=D}^{N-1} \lambda_k \mathbf{u}_k \mathbf{u}_k^H}_{\text{noise}} \quad (8.62)$$

The eigenvalues $\lambda_D, \ldots, \lambda_{N-1}$ are all equal to the noise power σ_n^2. We collect the signal and noise subspace vectors

$$\mathbf{U}_s \triangleq [\mathbf{u}_0, \ldots, \mathbf{u}_{D-1}] \quad (8.63)$$
$$\mathbf{U}_n \triangleq [\mathbf{u}_D, \ldots, \mathbf{u}_{N-1}] \quad (8.64)$$

and the signal eigenvalues

$$\mathbf{\Lambda}_s \triangleq \text{diag}\{\lambda_0, \ldots, \lambda_{D-1}\} \quad (8.65)$$

And the question now becomes how we estimate the D source angles $\boldsymbol{\psi}$ from the signal subspace defined by $\mathbf{U}_s$ and $\mathbf{\Lambda}_s$. A few useful identities can be proven from the subspace decomposition of $\mathbf{C_x}$ and its known structure. First, each of the signal vectors $\mathbf{v}(\psi_k)$ is contained in the subspace $\mathbf{U}_s$ and is orthogonal to the noise subspace $\mathbf{U}_n$.

$$\|\mathbf{v}^H(\psi_k)\mathbf{U}_s\|^2 = \sum_{j=0}^{D-1} |\mathbf{v}^H(\psi_k)\mathbf{u}_j|^2 = \sqrt{N} \quad (8.66)$$
$$\|\mathbf{v}^H(\psi_k)\mathbf{U}_n\|^2 = 0 \quad (8.67)$$
$$k = 0, 1, \ldots, D-1 \quad (8.68)$$

The solution, then, is to search over all of the possible pointing angles ψ, and find those that are not in the noise subspace $\mathbf{U}_n$

One of the most popular instantiations of a subspace methods is MUSIC [30, 31]. The Q function (an inverse of the power spectrum used in the beamformer approaches discussed previously) is given as

$$\widehat{Q}_{MUSIC}(\psi) = \mathbf{v}^H(\psi)\widehat{\mathbf{U}}_n\widehat{\mathbf{U}}_n^H\mathbf{v}(\psi) \quad (8.69)$$
$$\widehat{P}_{MUSIC}(\psi) = \frac{1}{\left|\widehat{Q}_{MUSIC}(\psi)\right|} \quad (8.70)$$

Equivalently (in terms of the signal and noise subspaces), this can be written with $\mathbf{U}_s$

$$\widehat{Q}_{MUSIC}(\psi) = \mathbf{v}^H(\psi) \left(\mathbf{I}_N - \widehat{\mathbf{U}}_s \widehat{\mathbf{U}}_s^H \right) \mathbf{v}(\psi) \tag{8.71}$$

This is achieved in MATLAB® via three stages. First, we take a sample covariance matrix (computed as was shown above for the MVDR beamformer), and perform an eigendecomposition. Note that MATLAB's eig command makes no sorting guarantee, so it must be sorted manually. Assuming that the dimensionality D is known, we collect all of the noise eigenvectors into the subspace matrix $\mathbf{U}_n$, and then evaluate (8.70).[5]

```
% Perform eigendecomposition, and sort the eigenvalues
[U,Lam] = eig(C);
lam = diag(Lam);
[~,idx_sort] = sort(abs(lam),'descend');
U_sort = U(:,idx_sort);
lam_sort = lam(idx_sort);

% Isolate Noise Subspace
Un = U_sort(:,D+1:end);
proj = Un*Un';

% Generate steering vectors
psi_vec = linspace(psi_min,psi_max,N_test_pts);
P = zeros(size(psi_vec));
for idx_pt = 1:numel(psi_vec)
    vv = v(psi_vec(idx_pt))/sqrt(N);
    Q = vv'*proj*vv;
    P(idx_pt) = 1./abs(Q);
end
```

This implementation is included in the accompanying MATLAB® package, under the array.music function.

```
P_music = array.music(x,v,2,psi_max,N_pts);
```

---

[5] If the dimensionality is not known, the provided example implementation estimates the noise power using the smallest eigenvalue, and sets a blind threshold at 3 dB above the estimated noise power. Any eigenvalues below that threshold are declared to be part of the noise subspace. This is not an optimal solution, and will fail at low SNR. For alternate approaches to estimating the number of sources, see [32–34] and countless more recent papers.

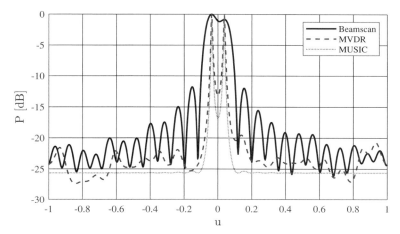

Figure 8.11 AOA estimation from the MUSIC algorithm, compared with standard and MVDR beamscan algorithms for two closely spaced sources with $\xi = 20$ dB, showing the improved separability of MVDR (when ξ is large).

We plot the estimation results in Figure 8.11 for the same simulation as before, but with the addition of a MUSIC estimator. MUSIC achieves similar resolution to the MVDR beamscan approach, but provides much more stable estimates at other angles, with a much more consistent -25-dB noise level.

Many other approaches exist, including variations and further developments on MUSIC, the *minimum-norm* algorithm [35, 36] (which uses a single vector in $\mathbf{U}_n$, rather than the full subspace, under the hypothesis that it is less likely to generate a false source position), and the *estimation of signal parameters via rotational invariance technique* (ESPRIT) algorithm [37, 38] (which does not require full knowledge of the steering vector $\mathbf{v}(\psi)$ and is, thus, robust to array perturbations).

The primary limitation of MUSIC and other subspace-based algorithms is that they are limited to $D \leq N$ sources. If $D > N$, then the subspace decomposition will be unable to isolate signal and noise subspaces. MUSIC, notably, also fails in the face of correlated signals, which can arise in cases of multipath.

8.4 PERFORMANCE ANALYSIS

In this section, we provide the CRLB on estimation performance for $\hat{\boldsymbol{\psi}}$ given M snapshots from an N-dimensional array. Recall that we provided the ML estimate for two signal models. In the first case, the signal $\mathbf{s}[m]$ is taken to be a complex Gaussian random variable with covariance matrix $\mathbf{C_s}$, and in the second it is a deterministic (but unknown) complex vector. We briefly discuss the CRLB for these two cases.

8.4.1 Gaussian Signal Model

The Gaussian signal model that we assume includes uncorrelated noise across the array ($\mathbf{n}[m] \sim \mathcal{CN}\left(\mathbf{0}_{N,1}, \sigma_n^2 \mathbf{I}_N\right)$), and unknown signal covariance matrix $\mathbf{C_s}$ that describes the correlation and power of each of the D sources. We express the covariance matrix of the received data vector $\mathbf{x}[m]$

$$\mathbf{C_x} = \mathbf{V}\mathbf{C_s}\mathbf{V}^H + \sigma_n^2 \mathbf{I}_N \tag{8.72}$$

In this case, the unknown parameter vector is

$$\boldsymbol{\vartheta} = \left[\boldsymbol{\psi}^T, \boldsymbol{\mu}^T, \sigma_n^2\right]^T \tag{8.73}$$

where the vector $\boldsymbol{\mu}$ contains all of the (real-valued) elements of $\mathbf{C_s}$

$$\boldsymbol{\mu} \triangleq \text{vec}\left[\mathbf{C_s}\right] \tag{8.74}$$

The derivation of this CRLB follows a known, but complex set of equations. The general process follows that of the receivers in Chapter 7, but the full derivation is omitted here. Interested readers are referred to Section 8.4.1 of [1] for a brief synopsis of the steps, or to [39] for the full details. The end result of this derivation is the bound

$$\mathbf{C}_{\boldsymbol{\psi}}(\mathbf{x}) \geq \frac{\sigma_n^2}{2M}\left[\Re\left\{\mathbf{K} \odot \mathbf{H}^T\right\}\right]^{-1} \tag{8.75}$$

where $\mathbf{K}$ and $\mathbf{H}$ are defined

$$\mathbf{K} \triangleq \mathbf{C_x}\left(\mathbf{I}_D + \frac{\mathbf{V}^H \mathbf{V}\mathbf{C_s}}{\sigma_n^2}\right)^{-1}\left(\frac{\mathbf{V}^H \mathbf{V}\mathbf{C_s}}{\sigma_n^2}\right) \tag{8.76}$$

$$\mathbf{H} \triangleq \mathbf{D}^H \mathbf{P}_\mathbf{V}^\perp \mathbf{D} \tag{8.77}$$

$$\mathbf{D} \triangleq \dot{\mathbf{V}} = \left[\frac{\partial \mathbf{v}(\psi_0)}{\partial \psi_0}, \ldots, \frac{\partial \mathbf{v}(\psi_{D-1})}{\partial \psi_{D-1}}\right] \tag{8.78}$$

Solving with the steering vector defined in (8.4), the individual gradients are defined as

$$\dot{\mathbf{v}}(\psi) \triangleq \frac{\partial \mathbf{v}(\psi)}{\partial \psi} = \left[0, \jmath \frac{2\pi d}{\lambda} \cos(\psi), \ldots, \jmath \frac{2\pi d}{\lambda}(N-1)\cos(\psi) \right] \odot v(\psi) \quad (8.79)$$

The function make_steering_vector in the enclosed MATLAB® package returns $\dot{\mathbf{v}}(\psi)$ as an optional second output

```
[v,v_dot] = array.make_steering_vector(d_lam,N);
```

Under a few conditions, namely that all the signals are well separated, the source covariance matrix $\mathbf{C_s}$ is not singular, and the signal power of each source is much larger than the noise, the CRLB approaches an asymptotic bound

$$\widetilde{\mathbf{C}}_\psi(\mathbf{x}) \geq \frac{\sigma_n^2}{2M} \left[\Re \left\{ \mathbf{C_s} \odot \mathbf{H}^T \right\} \right]^{-1} \quad (8.80)$$

This is sometimes referred to as the *stochastic* CRLB for array-based DF. It is included in the enclosed MATLAB® package as crlb_stochastic, and can be called with the signal covariance matrix Cs, noise power sigma_n2, true AOA vector psi_vec, number of snapshots M, and function handles for the steering vector and steering vector gradient v and v_dot, respectively.

```
crlb = array.crlb_stochastic(Cs,sigma_n2,psi_vec,M,v,v_dot);
```

Intuitively, this CRLB makes sense. The covariance matrix $\mathbf{C_s}$ contains the information in the source signals (how strong each is, and how correlated with each other they are), and the term $\mathbf{H}$ maps the ability of the array's steering vector to separate those signals.

8.4.2 Deterministic Signal Model

In the case of a deterministic signal, the parameter vector now has to expand to include both the real and imaginary components of $s[m]$ across all snapshots, rather than just the (real-valued) elements of the covariance matrix $\mathbf{C_s}$:

$$\boldsymbol{\vartheta} = \left[\boldsymbol{\psi}^T, \mathbf{s}_R^T, \mathbf{s}_I^T, \sigma_n^2 \right] \quad (8.81)$$

where the terms $\mathbf{s}_R$ and $\mathbf{s}_I$ are the collections of the real and imaginary components, respectively, of $\mathbf{s}[m]$ across all M snapshots.

$$\mathbf{s}_R = \Re\left\{\left[\mathbf{s}^T[0], \ldots, \mathbf{s}^T[M-1]\right]\right\} \tag{8.82}$$

$$\mathbf{s}_I = \Im\left\{\left[\mathbf{s}^T[0], \ldots, \mathbf{s}^T[M-1]\right]\right\} \tag{8.83}$$

Again, we leave out the details of this derivation, noting that it follows the same general approach as those in Chapter 8. Interested readers can find a step-by-step derivation in Section 8.4.4 of [1] (although this model for the parameter vector is slightly different, in that it interleaves the real and imaginary components of $\mathbf{s}[m]$ into a single, longer, parameter vector). A simpler form can also be seen in [40], which uses linearization to form a block diagonal Fisher information matrix, which is much simpler to invert.

The end result is that the CRLB in this case (referred to as the *conditional* CRLB—CCRB, by some and the *deterministic* CRLB by others) is given as

$$\mathbf{C}_{\psi}(\mathbf{x}) \geq \frac{\sigma_n^2}{2M} \left[\Re\left\{\mathbf{R}_\mathbf{s} \odot \mathbf{H}^T\right\}\right]^{-1} \tag{8.84}$$

where $\mathbf{H}$ is defined in (8.77), and

$$\mathbf{R}_\mathbf{s} = \frac{1}{M} \sum_{m=0}^{M-1} \mathbf{s}[m]\mathbf{s}^H[m] \tag{8.85}$$

is the sample spectral matrix of $\mathbf{s}$. If we repeat the assumptions of well separated signals, nonsingular covariance matrices, and high SNR, then the sample spectral matrix $\mathbf{R}_\mathbf{s}$ approaches the true spectral matrix $\mathbf{C}_\mathbf{s}$, and the CRLB asymptotically approaches

$$\widehat{\mathbf{C}}_{\psi}(\mathbf{x}) \geq \frac{\sigma_n^2}{2M} \left[\Re\left\{\mathbf{C}_\mathbf{s} \odot \mathbf{H}^T\right\}\right]^{-1} \tag{8.86}$$

This is implemented in the enclosed MATLAB® package

```
crlb = array.crlb_det(Cs,sigma_n2,psi_vec,M,v,v_dot);
```

The asymptotic CRLB is the same for both the Gaussian signal model in (8.80) and deterministic signal model (8.86). Thus, if SNR is sufficiently large and the sources are well separated, then it should not matter which signal model is used to estimate direction of arrival. The differences will arise in how the algorithms perform at low SNR.

8.4.2.1 SNR Notation

We define the SNR of each source's signal as the expected signal power (per temporal snapshot) divided by the noise, and collect those SNR values into a diagonal matrix $\boldsymbol{\Xi}$.

$$\xi_k = \frac{1}{M} \sum_{m=0}^{M-1} |s_k[m]|^2 \tag{8.87}$$

$$\boldsymbol{\Xi} = \text{diag}\{\xi_0, \ldots, \xi_{D-1}\} \tag{8.88}$$

If the sources are uncorrelated, then the covariance matrix $\mathbf{C_s}$ is diagonal, and can be expressed in terms of $\boldsymbol{\Xi}$ as

$$\mathbf{C_s} = \sigma_n^2 \boldsymbol{\Xi} \tag{8.89}$$

and we can rewrite the CRLB in terms of SNR, rather than the covariance matrix:

$$\widehat{\mathbf{C}}_{\boldsymbol{\psi}}(\mathbf{x}) \geq \frac{1}{2M} \left[\Re\left\{ \boldsymbol{\Xi} \odot \mathbf{H}^T \right\} \right]^{-1} \tag{8.90}$$

This representation is used in the provided CRLB function `array.crlb`, which is called with the SNR, true AOAs, number of snapshots, and the function handles to $\mathbf{v}(\psi)$ and $\partial \mathbf{v}(\psi)/\partial \psi$. If this is the notation available, then we call the CRLB function with the SNR (linear) and unit noise power

```
crlb = array.crlb_det(snr_vec_lin,1,psi_vec,M,v,dv);
```

In Figure 8.12, we plot both the stochastic and deterministic CRLB as a function of ξ, for a single source at $5°$ and array with $N = 11$ array elements and $M = 100$ temporal snapshots. We also plot the results of a Monte Carlo experiment for the beamscan and MUSIC solutions discussed above. We can see that, for very low SNR, the error saturates at roughly $50°$ and does not approach the CRLB until it drops below $1°$; this is consistent with the fact that the errors are only approximately Gaussian when they are small, and that the maximum possible angle error is $180°$. (We restrict the solutions to the forward hemisphere for array-based approaches.) We also note that, as ξ increases, the achieved results follow the stochastic CRLB and both converge toward the deterministic CRLB.

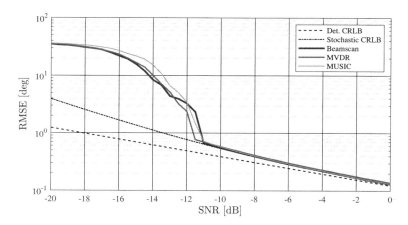

Figure 8.12 AOA performance based on both stochastic and deterministic CRLBs, and example implementations from this chapter, as a function of SNR.

8.4.2.2 Single Source

In the case of a single source, this becomes

$$\widehat{C}_\psi(\mathbf{x}) \geq \frac{1}{2M\xi H(\psi)} \quad (8.91)$$

where $H(\psi)$ is the scalar form of **H**, defined as

$$H(\psi) \triangleq \dot{\mathbf{v}}(\psi)^H \left(\mathbf{I}_N - \frac{\mathbf{v}(\psi)\mathbf{v}^H(\psi)}{\|\mathbf{v}(\psi)\|^2} \right) \dot{\mathbf{v}}(\psi) \quad (8.92)$$

With some algebraic manipulation (and realization that $H(\psi)$ is strictly real), the CRLB can be simplified as

$$\widehat{\mathbf{C}}_\psi(\mathbf{x}) \geq \frac{1}{2M\xi \left(\dot{\mathbf{v}}^H \dot{\mathbf{v}} - \frac{|\dot{\mathbf{v}}^H \mathbf{v}|^2}{\mathbf{v}^H \mathbf{v}} \right)} \quad (8.93)$$

which resembles the CRLB for amplitude-comparison in (7.24) from a number of steering angles (with the gain vector **g** replaced by the steering vector **v**). Further simplification reduces to the same CRLB as in the interferometer case (7.101) when

Table 8.2
Parameters for Example 8.2

(a) Datalink Transmitter

Parameter	Value
P_t	100W
G_t	0 dBi
B_s	500 kHz
f_0	950 MHz
L_t	3 dB
h_t	5 km

(b) Receiver

Parameter	Value
d/λ	.5
N	5 (array elements)
M	10 (samples)
G_r (per element)	0 dBi
L_r	2 dB
F_n	4 dB
B_n	1 MHz
h_r	10 m
T	1 μs

$N = 2$, $\alpha = 1$ and $\sigma_n^2 = \sigma_1^2 = \sigma_2^2$. This confirms that the array-based approach is merely an extension of the two-channel interferometer to N sensors and D sources. The execution of this simplification is left to the reader as an exercise.

Example 8.2 Airborne Datalink DF

At what range is the angular accuracy, as bounded by the CRLB, of an array-based AOA estimate better than $0.5°$ RMSE for sources near broadside ($\psi \approx 0$), given the source parameters in Table 8.2?

We begin by computing the received power as a function of range, using the same techniques from Chapter 2:

```
Pr_dB = 10*log10(Pt) + Gt - Lt + Gr - Lr - prop.pathLoss(R,...);
N_dB = 10*log10(utils.constants.kT*B_n)+F_n;
xi_dB = Pr_dB - N_dB;
xi_lin = 10.^(xi_dB/10);
```

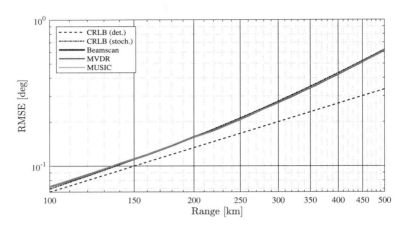

Figure 8.13 AOA estimation error as a function of standoff range for Example 8.2.

and compute the CRLB for each range. This can be done with a for loop or, more compactly, with the built-in arrayfun command.

```
C_psi_det = arrayfun(@(x) array.crlb_det(x,1,psi,M,v,v_dot),xi_lin);
C_th_det = (180/pi)^2*C_psi_det;
```

This process is repeated for the stochastic CRLB as well. They are plotted, along with a Monte Carlo experiment—for verification purposes—against range in Figure 8.13, which shows that the RMSE achieves $0.5°$ accuracy when the range to the target drops below ≈ 425 km.

8.5 PROBLEM SET

8.1 Create a steering vector for an array with $N = 31$ elements and $d = \lambda/2$, and another for $d = 3\lambda/4$. Plot the array factor.

8.2 Consider an array with $N = 11$ elements and $d = \lambda/2$. Plot the beamscan image for a superposition of two signals with amplitude $a = [1, 0.9]$ and AOA $\theta = [10°, 13°]$. Repeat with $N = 31$.

8.3 Add noise with variance $\sigma^2 = 0.25$ to the previous problem and replot the beamscan images.

8.4 Consider an $N = 30$ element array with $d = \lambda/2$. If the desired half-power beamwidth is $\theta_{\text{bw}} = 5°$ at broadside, what is the minimum achievable sidelobe level, given the common tapers in Table 8.1? What array length N is required to achieve this beamwidth with a Hann taper?

8.5 A sample data set is provided with the MATLAB® download, in the file hw/problem8_5.mat, which simulated three sources from a set of VHF push-to-talk radios. Process the data to determine the AOA for the three signals contained therein using the beam-scanning and MVDR beam-forming approaches. Comment on which method provides a better result.

8.6 A sample data set is provided with the MATLAB® download, in the file hw/problem8_6.mat, which simulated three sources from a set of VHF push-to-talk radios. Process the data to determine the AOA for the three signals contained therein using the beam-scanning, MVDR beam-forming, and MUSIC approaches. Comment on which method provides a better result.

8.7 At what range is the angular accuracy, as bounded by the CRLB, of an array based AOA estimate better than $.1°$ RMSE for sources near broadside ($\psi \approx 0$), given the source parameters in Table 8.2, with $N = 11$ and $M = 1$? Repeat for $d = \lambda$ (ignore the impact of ambiguities). Comment on the difference.

References

[1] H. L. Van Trees, *Optimum Array Processing: Part IV of Detection, Estimation and Modulation Theory*. Hoboken, NJ: Wiley-Interscience, 2002.

[2] M. A. Richards, J. Scheer, W. A. Holm, and W. L. Melvin, *Principles of Modern Radar*. Edison, NJ: SciTech Publishing, 2010.

[3] R. Mailloux, *Phased Array Antenna Handbook, Third Edition*. Norwood, MA: Artech House, 2017.

[4] M. Skolnik, *Radar Handbook, 3rd Edition*. New York, NY: McGraw-Hill Education, 2008.

[5] R. Willey, "Space tapaering of linear and planar arrays," *IRE Transactions on Antennas and Propagation*, vol. 10, no. 4, pp. 369–377, July 1962.

[6] J. Galejs, "Minimization of sidelobes in space tapered linear arrays," *IEEE Transactions on Antennas and Propagation*, vol. 12, no. 4, pp. 497–498, 1964.

[7] E. N. Gilbert and S. P. Morgan, "Optimum design of directive antenna arrays subject to random variations," *Bell System Technical Journal*, vol. 34, no. 3, pp. 637–663, 1955.

[8] Y. Lo, "A mathematical theory of antenna arrays with randomly spaced elements," *IEEE Transactions on Antennas and Propagation*, vol. 12, no. 3, pp. 257–268, May 1964.

[9] B. D. Steinberg, *Principles of aperture and array system design: Including random and adaptive arrays*. Hoboken, NJ: Wiley-Interscience, 1976.

[10] R. L. Haupt, "Thinned arrays using genetic algorithms," *IEEE Transactions on Antennas and Propagation*, vol. 42, no. 7, pp. 993–999, July 1994.

[11] M. G. Bray, D. H. Werner, D. W. Boeringer, and D. W. Machuga, "Optimization of thinned aperiodic linear phased arrays using genetic algorithms to reduce grating lobes during scanning," *IEEE Transactions on Antennas and Propagation*, vol. 50, no. 12, pp. 1732–1742, Dec 2002.

[12] L. Cen, W. Ser, W. Cen, and Z. L. Yu, "Linear sparse array synthesis via convex optimization," in *Circuits and Systems (ISCAS), Proceedings of 2010 IEEE International Symposium on*. IEEE, 2010, pp. 4233–4236.

[13] G. Oliveri and A. Massa, "Bayesian compressive sampling for pattern synthesis with maximally sparse non-uniform linear arrays," *IEEE Transactions on Antennas and Propagation*, vol. 59, no. 2, pp. 467–481, 2011.

[14] D. L. Donoho, "Compressed sensing," *IEEE Transactions on Information Theory*, vol. 52, no. 4, pp. 1289–1306, April 2006.

[15] L. Carin, D. Liu, and B. Guo, "Coherence, compressive sensing, and random sensor arrays," *IEEE Antennas and Propagation Magazine*, vol. 53, no. 4, pp. 28–39, Aug 2011.

[16] E. Fishler, A. Haimovich, R. Blum, D. Chizhik, L. Cimini, and R. Valenzuela, "Mimo radar: an idea whose time has come," in *Proceedings of the 2004 IEEE Radar Conference (IEEE Cat. No.04CH37509)*, April 2004, pp. 71–78.

[17] J. Kantor and S. K. Davis, "Airborne gmti using mimo techniques," in *2010 IEEE Radar Conference*, May 2010, pp. 1344–1349.

[18] A. J. Paulraj, D. A. Gore, R. U. Nabar, and H. Bolcskei, "An overview of mimo communications-a key to gigabit wireless," *Proceedings of the IEEE*, vol. 92, no. 2, pp. 198–218, 2004.

[19] B. D. Van Veen and K. M. Buckley, "Beamforming: A versatile approach to spatial filtering," *IEEE assp magazine*, vol. 5, no. 2, pp. 4–24, 1988.

[20] S. A. Schelkunoff, "A mathematical theory of linear arrays," *The Bell System Technical Journal*, vol. 22, no. 1, pp. 80–107, Jan 1943.

[21] T. T. Taylor, "Design of line-source antennas for narrow beamwidth and low side lobes," *Transactions of the IRE Professional Group on Antennas and Propagation*, vol. 3, no. 1, pp. 16–28, Jan 1955.

References

[22] E. T. Bayliss, "Design of monopulse antenna difference patterns with low sidelobes*," *Bell System Technical Journal*, vol. 47, no. 5, pp. 623–650, 1968.

[23] F. R. Hill and R. L. Pickholtz, "Estimating the number of signals using the eigenvalues of the correlation matrix," in *IEEE Military Communications Conference, 'Bridging the Gap. Interoperability, Survivability, Security'*, Oct 1989, pp. 353–358 vol.2.

[24] P. Stoica and M. Cedervall, "An Eigenvalue-Based Detection Test for Array Signal Processing in Unknown Correlated Noise Fields," *IFAC Proceedings Volumes*, vol. 29, no. 1, pp. 4098–4103, 1996.

[25] P. Stoica and A. Nehorai, "Performance study of conditional and unconditional direction-of-arrival estimation," *IEEE Transactions on Acoustics, Speech, and Signal Processing (ICASSP)*, vol. 38, no. 10, pp. 1783–1795, 1990.

[26] R. Poisel, *Electronic warfare target location methods*. Norwood, MA: Artech House, 2012.

[27] S. Boyd and L. Vandenberghe, *Convex optimization*. Cambridge, UK: Cambridge University Press, 2004.

[28] J. Capon, "High-resolution frequency-wavenumber spectrum analysis," *Proceedings of the IEEE*, vol. 57, no. 8, pp. 1408–1418, 1969.

[29] P. Stoica, "On robust capon beamforming and diagonal loading," *IEEE Transactions on Signal Processing*, vol. 51, no. 7, pp. 1702–1715, July 2003.

[30] R. Schmidt, "Multiple Emitter Location and Signal Parameter Estimation," Ph.D. dissertation, Stanford University, 1981.

[31] ——, "Multiple emitter location and signal parameter estimation," *IEEE transactions on antennas and propagation*, vol. 34, no. 3, pp. 276–280, 1986.

[32] M. Wax and I. Ziskind, "Detection of the number of coherent signals by the mdl principle," *IEEE Transactions on Acoustics, Speech, and Signal Processing*, vol. 37, no. 8, pp. 1190–1196, 1989.

[33] W. Chen, K. M. Wong, and J. P. Reilly, "Detection of the number of signals: A predicted eigenthreshold approach," *IEEE Transactions on Signal Processing*, vol. 39, no. 5, pp. 1088–1098, 1991.

[34] J.-F. Gu, P. Wei, and H.-M. Tai, "Detection of the number of sources at low signal-to-noise ratio," *IET Signal Processing*, vol. 1, no. 1, pp. 2–8, 2007.

[35] S. Reddi, "Multiple source location-a digital approach," *IEEE Transactions on Aerospace and Electronic Systems*, no. 1, pp. 95–105, 1979.

[36] R. Kumaresan and D. W. Tufts, "Estimating the angles of arrival of multiple plane waves," *IEEE Transactions on Aerospace and Electronic Systems*, no. 1, pp. 134–139, 1983.

[37] R. Roy, "ESPRIT: Estimation of Signal Parameters via Rotational Invariance Technique," Ph.D. dissertation, Stanford University, 1987.

[38] R. Roy and T. Kailath, "ESPRIT-estimation of signal parameters via rotational invariance techniques," *IEEE Transactions on Acoustics, Speech, and Signal Processing*, vol. 37, no. 7, 1989.

[39] A. J. Weiss and B. Friedlander, "On the cramer-rao bound for direction finding of correlated signals," *IEEE Transactions on Signal Processing*, vol. 41, no. 1, pp. 495–, January 1993.

[40] P. Stoica and E. G. Larsson, "Comments on "linearization method for finding cramer-rao bounds in signal processing" [with reply]," *IEEE Transactions on Signal Processing*, vol. 49, no. 12, 2001.

Part III

Geolocation of Threat Emitters

Chapter 9

Geolocation of Emitters

Part II discussed general estimation theory and the use of direction-finding and array-processing methods to determine the angle of incoming signals. In many cases, knowledge of the physical location from which a signal is emitted can allow for a greater variety of responses, in addition to the increased situational awareness that comes from knowing precisely where the threat is located. Some examples include the ability to cue a kinetic response, to alert friendly forces to the location of the threat, and to build an estimate of the adversary's force disposition.

In this chapter, we briefly introduce the concept of geolocation, and describe common measures of performance. In the remaining chapters, we will describe several approaches for geolocation. The first, triangulation, is an extension of direction finding to multiple receiver sites. *Time difference of arrival* (TDOA) and *frequency difference of arrival* (FDOA) are precision geolocation techniques that rely on processing differences in the signals received at a constellation of sensor sites. These three approaches are discussed in Chapters 10, 11, and 12, respectively. A hybrid technique that combines estimation from TDOA and FDOA is described in Chapter 13.

9.1 BACKGROUND

Geolocation is the process of estimating the physical coordinates ($\mathbf{x}$) from which a signal was transmitted. The estimated position is noted here with $\widehat{\mathbf{x}}$.

The signal $p(t)$ is transmitted from an unknown source position $\mathbf{x}$ with velocity $\mathbf{v}$ and propagates to the sensor position $\mathbf{x}_i$, which has velocity $\mathbf{v}_i$, where it is received as signal $y_i(t)$. Under some basic assumptions (constant propagation velocity, no diffraction or complex propagation paths), the received signal is written as

$$s(t) = p(t - \tau_i)e^{j2\pi \frac{v_i}{\lambda}} \tag{9.1}$$

$$\tau_i = \frac{\|\mathbf{x} - \mathbf{x}_i\|}{c} \tag{9.2}$$

$$v_i = (\mathbf{v} + \mathbf{v}_i)^T \frac{(\mathbf{x} - \mathbf{x}_i)}{\|\mathbf{x} - \mathbf{x}_i\|} \tag{9.3}$$

where τ_i is the time delay between $\mathbf{x}$ and $\mathbf{x}_i$, and v_i is the projection of relative velocity onto the line of sight between the transmitter and receiver, and c is the speed of light.

All of the geolocation techniques use will involve either an extension of AOA techniques from Chapters 7 and 8, or estimation of these delay and doppler terms.

9.2 PERFORMANCE METRICS

There are a number of potential performance criteria that can be used to analyze geolocation algorithms, or fielded emitter location systems. In general, the performance metrics are a function of the bias term—$\mathbf{b}$—and covariance matrix—$\mathbf{C}_{\hat{\mathbf{x}}}$—of errors between the estimated and true positions

$$\mathbf{b} = E\{(\hat{\mathbf{x}} - \mathbf{x})\} \tag{9.4}$$

$$\mathbf{C}_{\hat{\mathbf{x}}} = E\left\{(\hat{\mathbf{x}} - \mathbf{x} - \mathbf{b})(\hat{\mathbf{x}} - \mathbf{x} - \mathbf{b})^T\right\} \tag{9.5}$$

where $E\{\cdot\}$ is the expectation operator. For a three-dimensional system, there are nine entries in the error covariance matrix $\mathbf{C}_{\hat{\mathbf{x}}}$, the noise within each of the three cardinal dimensions, along with cross-covariance terms that define how they vary together.

9.2.1 Error Ellipse

The most straightforward measure of position accuracy is the error ellipse. For any two dimensions, the error ellipse can be used to trace the distribution of estimates about the bias point. It is computed simply from the ellipse equation

$$\left(\frac{x - b_x}{\sigma_x}\right)^2 + \left(\frac{y - b_y}{\sigma_y}\right)^2 = \gamma \tag{9.6}$$

Table 9.1
Error Ellipse Scale Factor and Confidence Intervals

Confidence Interval	Scale Factor γ
39.35%	1.000
50%	1.386
63.2%	2.000
75%	2.773
90%	4.601
95%	5.991

where b_x and b_y are the x and y components of the bias vector $\mathbf{b}$, σ_x and σ_y are the x and y variance terms from $\mathbf{C}_{\hat{\mathbf{x}}}$, respectively, and γ is a scale parameter used to determine the area of the ellipse. The selection of γ controls what percentage of estimates fall within the ellipse; thereby defining its confidence interval.

To compute γ, we define the error term $\hat{z}$, which is the scale parameter of the ellipse on which a given estimate sits. $\hat{z}$ is distributed as a chi-squared random variable with two degrees of freedom [1]. The probability that $\hat{z} < \gamma$ is provide the probability that a given estimate lies within the error ellipse with scale parameter γ

$$\hat{z} = \left(\frac{\hat{x}-b_x}{\sigma_x}\right)^2 + \left(\frac{\hat{y}-b_y}{\sigma_y}\right)^2 \tag{9.7}$$

$$\mathcal{P}\{\hat{z} < \gamma\} = 1 - e^{-\gamma/2} \tag{9.8}$$

where $\hat{x}$ and $\hat{y}$ are the x and y components of the estimated position $\hat{\mathbf{x}}$. Table 9.1 provides an evaluation of (9.8) for several representative values.

Equation (9.6) is only relevant if the errors in the two dimensions used (here x and y) are uncorrelated, that is the covariance matrix of the error in those two dimensions is diagonal. If there is any correlation, then the standard deviations σ_x and σ_y should be replaced with the square root of the eigenvalues of the covariance matrix, $\sqrt{\lambda_1}$ and $\sqrt{\lambda_2}$, as those eigenvalues define the semimajor and semiminor axis of the error ellipse. The ellipse must then be rotated to align with the eigenvectors. The variables x and y should be replaced with dummy variables

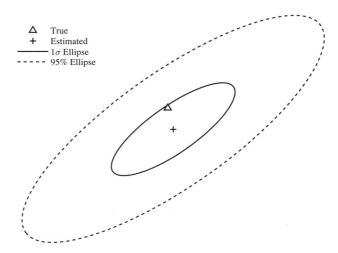

Figure 9.1 Drawing of a 2-D error ellipse (true location marked with a △, expected estimate marked with a +). The 1-σ error ellipse is marked with a solid line, and its corresponding eigenvectors with dotted lines from the biased position estimate. The dashed line depicts a 95% confidence interval.

$\widetilde{x}_1$ and $\widetilde{x}_2$, respectively. The x and y values can then be computed with the rotation

$$\alpha = \tan^{-1}(v_y/v_x) \tag{9.9}$$
$$x = \cos(\alpha)\widetilde{x}_1 - \sin(\alpha)\widetilde{x}_2 \tag{9.10}$$
$$y = \sin(\alpha)\widetilde{x}_1 + \cos(\alpha)\widetilde{x}_2 \tag{9.11}$$

where **v** is the eigenvector associated with the first eigenvalue, and v_x and v_y are the x and y components of **v**, respectively. This relationship is illustrated in Figure 9.1 for a sample case; the true target position is marked with a '+', the bias point is shown with a 'o', the 1-σ error ellipse (39.35% confidence interval) is traced with a solid line, and the 95% confidence interval with a dashed line. The two arcs trace the semimajor and semiminor ellipses. In this case, there is correlation between x and y errors, so the eigenvalues of the covariance matrix are used.

In three dimensions, the error ellipse is replaced by an error ellipsoid, and the equations here must be reformulated. Typically 3-D errors are represented by their projections onto the various 2-D planes (x-y, y-z, and x-z).

Example 9.1 Computing the Error Ellipse

Given the covariance matrix

$$\mathbf{C} = \begin{bmatrix} \sigma_x^2 & \sigma_{xy} \\ \sigma_{yx} & \sigma_y^2 \end{bmatrix} = \begin{bmatrix} 10 & 3 \\ 3 & 5 \end{bmatrix}, \tag{9.12}$$

find the equation for the error ellipse that describes a 90% confidence interval. Assume the estimate is unbiased ($\mathbf{b} = \mathbf{0}$).

The first step is to find the two eigenvectors and their corresponding eigenvalues. Manual computation methods can be found in Chapter 2 of [2], among other sources. To do so in MATLAB, one must simply call the eig command.

```
> [V,Lam] = eig(C)
V =
    0.4242   -0.9056
   -0.9056   -0.4242
Lam =
    3.5949        0
         0  11.4051
```

This means that the eigenvalues are $\lambda_1 = 3.5949$ and $\lambda_2 = 11.4051$, and their corresponding eigenvectors are $v_1 = [0.4242, -0.9056]$ and $v_2 = [-0.9056, -0.4242]$. λ_2 is the dominant eigenvalue.

The next step is to look up the proper scale factor from Table 9.1, which results in $\gamma = 4.601$, and construct the error ellipse using the two eigenvalues and dummy variables $\tilde{x}$ and $\tilde{y}$.

$$4.601 = \left(\frac{\tilde{x}_1}{3.3771}\right)^2 + \left(\frac{\tilde{x}_2}{1.8960}\right)^2 \tag{9.13}$$

where the 3.3771 is the square root of the dominant eigenvalue, and 1.8960 is the square root of the smaller eigenvalue. Next, we compute the rotation angle with the y and x components of the dominant eigenvector

```
> v_max = V(:,2);
> lam_max = lam(2,2);
> alpha = atan2(v_max(2),v_max(1))
alpha =
   -2.7036
```

For reference, that is 154 degrees. Finally, using the rotation angle and (9.10) and (9.11), we compute the ellipse in terms of the x and y coordinates. The approach to draw the ellipse taken here is to note that the radius along the semiminor and semimajor axes is given by:

$$r_1^2 = \gamma \lambda_{\max} \qquad (9.14)$$
$$r_2^2 = \gamma \lambda_{\min} \qquad (9.15)$$

We then take a series of angular positions $\theta \in [0, 2\pi]$ and compute the ellipse coordinates

$$\widetilde{x}_1 = r_1 \cos(\theta) \qquad (9.16)$$
$$\widetilde{x}_2 = r_2 \sin(\theta) \qquad (9.17)$$

Once the ellipse is given in these coordinates, we rotate them using the expressions in (9.10) and (9.11). In MATLAB, this can be written with the following script:

```
theta = linspace(0,2*pi,361);
r1 = sqrt(gamma*lam_max);
r2 = sqrt(gamm*lam_min);
x1 = r1*cos(theta);
x2 = r2*sin(theta);
x = x1*cos(alpha) +x2*sin(alpha);
x = -x1*sin(alpha)+x2*cos(alpha);
```

This is plotted below in Figure 9.2. We can also use the included utility utils.drawErrorEllipse to draw the ellipse.

```
utils.drawErrorEllipse(x,C,numPts,confInterval);
```

where x is the center point of the ellipse (true position plus bias), C is the covariance matrix as defined above, numPts is the number of points to draw (100 to 1,000 typically works well), and confInterval is the desired confidence interval (between 0 and 1).

9.2.2 CEP

CEP is defined simply as the radius of a circle—centered on the emitter location, not the bias location—such that a given percentage of estimates fall within that circle. The most commonly used metrics are CEP_{50}—wherein 50% of the estimates fall

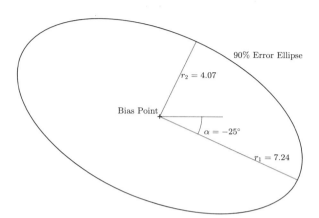

Figure 9.2 Error ellipse from Example 9.1. The + is the bias point, while the solid line indicates the 95% confidence interval error ellipse. The dotted lines show the semimajor and semiminor axes r_1 and r_2, respectively.

within the circle—and CEP_{90}—wherein 90% of the estimates fall within the circle. This relationship is shown in Figure 9.3 for CEP_{50}, and a 50% error ellipse. CEP is very useful for weapons delivery and seeker handoff, since it defines the probability that a target location estimate falls within a circular window of a certain radius, and hence is a meaningful metric for geolocation accuracy.

The definition of the CEP is rather straightforward, but its computation is not; papers studying the CEP go back to the 1950s [3, 4]. The value CEP is the radius of a circle that encloses errors with a given probability. If we define the total error $\widetilde{z}$

$$\widetilde{z} = \sqrt{\widehat{x}^2 + \widehat{y}^2} \tag{9.18}$$

then CEP can be defined as

$$\int_0^{CEP_\gamma} f_z(\widetilde{z})\, d\widetilde{z} = \frac{\gamma}{100} \tag{9.19}$$

The error term $\widetilde{z}$ is the square root of the weighted sum of noncentral chi-squared random variables with one degree of freedom (since both $\widehat{x}$ and $\widehat{y}$ are Gaussian random variables with nonzero mean) [1]. This quantity does not exist in closed form, although several approximations have been made.

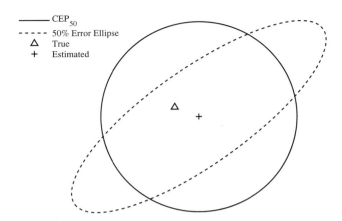

Figure 9.3 Illustration of the CEP calculation for a biased estimator with the true position ($\triangle$), estimated position $\hat{\mathbf{x}}_0$ (+), CEP (solid line), and error ellipse (dashed line).

The most popular approximation is given for CEP_{50} [5, 6]

$$\text{CEP}_{50} = \begin{cases} .59\,(\sigma_s + \sigma_l) & \frac{\sigma_s}{\sigma_l} \geq .5 \\ \sigma_l\left(.67 + .8\frac{\sigma_s^2}{\sigma_l^2}\right) & \frac{\sigma_s}{\sigma_l} < .5 \end{cases} \qquad (9.20)$$

where σ_s is the smaller of the two error terms and σ_l is the larger. The first approximation is valid when the error ellipse has low eccentricity (the two error terms are approximately equal), and the latter is used when one error term dominates. This approximation is accurate to 1%, provided that the estimate $\hat{\mathbf{x}}$ is unbiased and that the errors are independent in both x and y. The latter assumption can be alleviated in the same manner as with the error ellipse, by replacing the standard deviation terms σ with the square roots of the eigenvalues of the error covariance matrix **R**.

Just as with the error ellipse, CEP is typically computed in two dimensions and can be done for (x, z) or (y, z) in addition to the (x, y) variant shown here. However, since CEP is scalar, there is also the opportunity to include all three error terms; the result is *spherical error probable* (SEP). There are many approximations that can be used, based on the relative value of the standard deviation of error in the three dimensions, many of which are presented in [7]. A simple approximation to SEP, for 50 and 90 percent, respectively, is given [8]

$$SEP_{50} \approx .51\,(\sigma_x + \sigma_y + \sigma_z) \qquad (9.21)$$
$$SEP_{90} \approx .833\,(\sigma_x + \sigma_y + \sigma_z) \qquad (9.22)$$

These approximations also assume that the $\hat{x}$ is unbiased. Additional approximations were reported in [9], derived through curve fitting, and reported to have accuracy of roughly 2% across a broad range of error variance ratios between the spatial dimensions. In addition, there is a MATLAB® script available on the Mathworks Central File Exchange that mathematically computes the SEP, given a set of error terms [10].

Example 9.2 Compute the CEP

For the same covariance matrix defined above in Example 9.1, compute the CEP_{50}. Just as in Example 9.1, we start by computing the eigenvalues, which are then given as $\lambda_{max} = 11.4051$ and $\lambda_{min} = 3.4959$. To compute CEP, we next take the square root of the ratio:

$$\sqrt{\frac{\lambda_{min}}{\lambda_{max}}} = \frac{1.8960}{3.3771} = 0.56 \qquad (9.23)$$

Since this is greater than 0.5, the equation from (9.20) to use is

$$CEP_{50} = .59\left(\sqrt{\lambda_{min}} + \sqrt{\lambda_{max}}\right) = 3.112 \qquad (9.24)$$

This result is plotted below in Figure 9.4 for the CEP (solid line) and 50% error ellipse (dashed line).

9.2.3 MATLAB® Code

The provided MATLAB® code contains functions to compute each of the performance metrics listed here, contained within the functions utils.draw_errorEllipse, utils.compute_cep50, and utils.draw_cep50.

9.3 CRLB

The concept of statistical bounds for estimation performance is first introduced in Chapter 6. We briefly discuss here the application of some of these bounds to geolocation estimates.

Since the CRLB computes an upper bound on the error covariance matrix $C_{\hat{x}}$ for an unbiased estimate $\hat{x}$, it can be used to compute corresponding bounds on the measures already introduced, such as the CEP_{50} or RMSE. Application of the CRLB to geolocation algorithms is well studied [11].

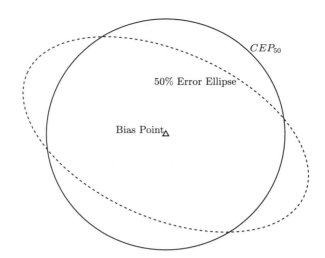

Figure 9.4 CEP and error ellipse for the covariance matrix in Example 9.2. The triangle shows the estimator bias point, the dashed line is the 50% error ellipse (for reference) and the solid line is the circle with radius given by CEP_{50}.

Recall from Chapter 6 that the CRLB is a lower bound on the elements of the covariance matrix $C_{\hat{x}}$, and is given by the inverse of the Fisher information matrix (FIM). In the general Gaussian case, with a covariance matrix C on the input data vector, the FIM is given

$$\mathbf{F} = \mathbf{J}\mathbf{C}^{-1}\mathbf{J}^T, \tag{9.25}$$

where $\mathbf{J}$ is the Jacobian matrix of the received data vector, defined

$$\mathbf{J} = \nabla_{\mathbf{x}}\mathbf{f}(\mathbf{x}) = [\nabla_{\mathbf{x}}f_1(\mathbf{x}), \ldots, \nabla_{\mathbf{x}}f_{L-1}(\mathbf{x})] \tag{9.26}$$

where $\mathbf{f}(\mathbf{x})$ is the vector of L received signals given an emitter at location $\mathbf{x}$, and $\nabla_{\mathbf{x}}f_l(\mathbf{x})$ is the gradient of the lth received signal at that location (an 2x1 or 3x1 data vector, for 2-D or 3-D solutions, respectively).

The structure of the Jacobian is dependent on the equations used to estimate emitter location and the specific geometry of the emitters in relation to the sensors. The measurement covariance matrix $\mathbf{C}$ is a function not only of sensor performance and the parameter being estimated, but also the achieved signal to noise ratio (which is dependent on the distance from the emitter to each receiver).

The dependence of these matrices on geometry means that the CRLB is difficult to fully solve analytically, and often must be evaluated numerically for specific situations. Insight is typically gained by positing a geometry and plotting changes in the CRLB as some key parameter is varied.

9.4 TRACKERS

The geolocation solutions in the following chapters are designed and presented as *single-snapshot* solutions (e.g., given a set of angle, time, or frequency measurements, we estimate the source's position). In many cases, this process will occur repeatedly. Thus, it is often advisable to run a tracker, as was discussed briefly in Chapter 6.

Trackers, when well designed, consider information from prior measurements and position estimates, as well as motion models, to improve measurement accuracy over time. Another benefit of trackers is that they provide a convenient initial estimate for the iterative solutions presented in the following chapters. At each time step, the iterative geolocation solution can be initialized with the previous time step's solution (or the prediction that was generated for the current time step, as is done in Kalman filters).

Although trackers are not discussed, interested readers are referred to early work on tracking with bearing only measurements from one or more sensors [12, 13], and with TDOA and FDOA measurements [14].

9.5 GEOLOCATION ALGORITHMS

The rest of Part III is organized as follows. Chapter 10 introduces the concept of *trilateration*, which is the process of combining multiple DF estimates to form a position estimate. Chapter 11 discusses TDOA, which is sometimes referred to as a *hyperbolic* geolocation technique. Chapter 12 introduces a similar technique, FDOA, that works on estimated Doppler shifts between sensors, rather than time delays. Finally, Chapter 13 discusses a hybrid TDOA/FDOA technique.

9.5.1 Ongoing Research

The algorithms discussed here represent three classical approaches to passively locating sources and include modern developments. In many cases, the algorithms (as presented) rely on a set of classic assumptions, such as a single target in isolation, or perfect knowledge of sensor position, and uncorrelated noise. There is a rich field of ongoing research into many of these areas, including extensions of the classical algorithms to more general problems, such as moving sensors [15, 16], sensor position or other system calibration errors [17, 18], or multiple sources [19–21]. Some research is conducted on occluded positioning, particularly for indoor and urban environments, or through walls [22–25]. These sources are not meant to be exhaustive or definitive, but rather reflect some of the breadth of recent studies in the field.

9.6 PROBLEM SET

9.1 Find the error ellipse for a 90% confidence interval, assuming an unbiased estimate, with covariance matrix

$$\mathbf{C} = \begin{bmatrix} 8 & 4 \\ 4 & 6 \end{bmatrix}$$

9.2 Find the error ellipse for a 50% confidence interval, assuming an unbiased estimate, with covariance matrix

$$\mathbf{C} = \begin{bmatrix} 5 & 0 \\ 0 & 7 \end{bmatrix}$$

9.3 Find the error ellipse for a 95% confidence interval, assuming an unbiased estimate, with covariance matrix

$$\mathbf{C} = \begin{bmatrix} 2 & 2 \\ 2 & 7 \end{bmatrix}$$

9.4 Compute the CEP_{50} for the covariance matrix in Problem 9.2.

9.5 Compute the CEP_{50} for the covariance matrix in Problem 9.3.

9.6 Compute the RMSE for the covariance matrix in Problem 9.2.

9.7 Compute the RMSE for the covariance matrix in Problem 9.3.

References

[1] M. K. Simon, *Probability distributions involving Gaussian random variables: A handbook for engineers and scientists.* New York, NY: Springer Science & Business Media, 2007.

[2] L. L. Scharf, *Statistical signal processing.* Reading, MA: Addison-Wesley, 1991.

[3] R. L. Elder, "An examination of circular error probable approximation techniques," Air Force Institute of Technology, School of Engineering, Tech. Rep., 1986.

[4] RAND Corporation, "Offset circle probabilities," RAND Corporation, Tech. Rep., 1952.

[5] L. H. Wegner, "On the accuracy analysis of airborne techniques for passively locating electromagnetic emitters," RAND, Tech. Rep., 1971.

[6] W. Nelson, "Use of circular error probability in target detection," MITRE, Tech. Rep., 1988.

[7] R. J. Schulte and D. W. Dickinson, "Four methods of solving for the spherical error probably associated with a three-dimensional normal distribution," Air Force Missile Development Center, Tech. Rep., 1968.

[8] National Research Council, *The global positioning system: A shared national asset.* Washington, D.C.: National Academies Press, 1995.

[9] D. R. Childs, D. M. Coffey, and S. P. Travis, "Error statistics for normal random variables," Naval Underwater Systems Center, Tech. Rep., 1975.

[10] M. Kleder, "Sep - an algorithm for converting covariance to spherical error probable," 2004, *. [Online]. Available: https://www.mathworks.com/matlabcentral/fileexchange/5688-sep-an-algorithm-for-converting-covariance-to-spherical-error-probable

[11] A. Yeredor and E. Angel, "Joint TDOA and FDOA estimation: A conditional bound and its use for optimally weighted localization," *IEEE Transactions on Signal Processing*, vol. 59, no. 4, pp. 1612–1623, April 2011.

[12] S. Fagerlund, "Target tracking based on bearing only measurements," MIT Laboratory for Information and Decision Systems, Tech. Rep. ADA100758, 1980.

[13] B. La Scala and M. Morelande, "An analysis of the single sensor bearings-only tracking problem," in *2008 11th International Conference on Information Fusion.* IEEE, 2008, pp. 1–6.

[14] D. Mušicki, R. Kaune, and W. Koch, "Mobile emitter geolocation and tracking using TDOA and FDOA measurements," *IEEE transactions on signal processing*, vol. 58, no. 3, pp. 1863–1874, 2010.

[15] K. C. H. and, "An accurate algebraic solution for moving source location using TDOA and FDOA measurements," *IEEE Transactions on Signal Processing*, vol. 52, no. 9, pp. 2453–2463, Sep. 2004.

[16] F. Fletcher, B. Ristic, and Darko Mušicki, "Recursive estimation of emitter location using TDOA measurements from two UAVs," in *2007 10th International Conference on Information Fusion*, July 2007, pp. 1–8.

[17] K. Ho, X. Lu, and L. Kovavisaruch, "Source Localization Using TDOA and FDOA Measurements in the Presence of Receiver Location Errors: Analysis and Solution," *Trans. Sig. Proc.*, vol. 55, no. 2, pp. 684–696, Feb. 2007.

[18] Z. Aliyazicioglu, H. Hwang, M. Grice, and A. Yakovlev, "Sensitivity analysis for direction of arrival estimation using a root-music algorithm." *Engineering Letters*, vol. 16, no. 3, 2008.

[19] C. Blandin, A. Ozerov, and E. Vincent, "Multi-source TDOA estimation in reverberant audio using angular spectra and clustering," *Signal Processing*, vol. 92, no. 8, pp. 1950 – 1960, 2012, latent Variable Analysis and Signal Separation.

[20] J. Scheuing and B. Yang, "Disambiguation of TDOA estimation for multiple sources in reverberant environments," *IEEE Transactions on Audio, Speech, and Language Processing*, vol. 16, no. 8, pp. 1479–1489, Nov 2008.

[21] M. R. Azimi-Sadjadi, A. Pezeshki, and N. Roseveare, "Wideband doa estimation algorithms for multiple moving sources using unattended acoustic sensors," *IEEE Transactions on Aerospace and Electronic Systems*, vol. 44, no. 4, pp. 1585–1599, 2008.

[22] J. H. DiBiase, H. F. Silverman, and M. S. Brandstein, "Robust localization in reverberant rooms," in *Microphone Arrays*, M. Brandstein and D. Ward, Eds. Berlin: Springer, 2001.

[23] K. Chetty, G. E. Smith, and K. Woodbridge, "Through-the-wall sensing of personnel using passive bistatic wifi radar at standoff distances," *IEEE Transactions on Geoscience and Remote Sensing*, vol. 50, no. 4, pp. 1218–1226, 2012.

[24] K. Witrisal, P. Meissner, E. Leitinger, Y. Shen, C. Gustafson, F. Tufvesson, K. Haneda, D. Dardari, A. F. Molisch, A. Conti, and M. Z. Win, "High-accuracy localization for assisted living: 5g systems will turn multipath channels from foe to friend," *IEEE Signal Processing Magazine*, vol. 33, no. 2, pp. 59–70, March 2016.

[25] A. O'Connor, P. Setlur, and N. Devroye, "Single-sensor rf emitter localization based on multipath exploitation," *IEEE Transactions on Aerospace and Electronic Systems*, vol. 51, no. 3, pp. 1635–1651, July 2015.

Chapter 10

Triangulation of AOA Measurements

10.1 BACKGROUND

Triangulation is the use of multiple bearing measurements to determine a source's position. In a perfect world, two measurements are sufficient for localization.[1] However, the presence of noise means that the use of additional bearing estimates will improve accuracy (assuming the errors in each bearing estimate are at least partially uncorrelated).

In this chapter, we consider a number of solutions to the problem of solving for source position given N noisy bearing estimates $\widehat{\psi}_i$, $i = 0, \ldots, N-1$. We first begin with geometric formulation of the problem, then discuss several solutions, including deterministic solutions posed from the geometry of the problem, the maximum likelihood estimate, and two iterative solutions. There is a rich variety of solutions in the literature, but the methods presented here form a basis for understanding those results.

Recall from Chapter 9 that, in this section of the text, we use x to represent geographic coordinates in 2-D or 3-D, while earlier chapters used it to represent a sample vector.

References for triangulation are plentiful. Chapter 7 of [1] contains a brief introduction and description of triangulation and of the impact of geometry on errors, as does Section 17.3 of [2]. Those references are useful for an operational context on requirements and uses of triangulation. A more detailed description of solutions and error performance can be found in Chapter 2 of [3], and with

---

1 For 2-D localization, a pair of azimuth measurements is sufficient; for 3-D localization, both measurements must include either azimuth and elevation angles or rotation and cone angles, as discussed briefly in Chapter 8.

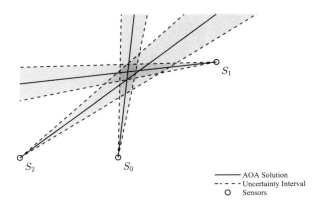

Figure 10.1 Illustration of AOA estimates from three sensors, with uncertainty intervals. Solutions for three or more sensors have a near-zero chance of intersecting at a common point.

a satellite-based perspective in Section 7.2 of [4], both of which are suitable for theoretical analysis and algorithm development. Early formulations date back to the 1940s [5].

10.2 FORMULATION

Consider the 2-D geometry in Figure 10.2, in which we have three sensors and a target; the line of bearing (relative to the +x direction) is given as ψ_i, $i = 0, \ldots, N-1$. In each case, the angle can be related to the position of the source $\mathbf{x} = (x, y)$ and the sensor $\mathbf{x}_i = (x_i, y_i)$

$$\psi_i = \tan^{-1}\left(\frac{y - y_i}{x - x_i}\right) \tag{10.1}$$

We collect the N available noisy bearing measurements into a system of equations

$$\boldsymbol{\psi} = \mathbf{p}(\mathbf{x}) + \mathbf{n} \tag{10.2}$$

where the AOA vector $\mathbf{p}$ is defined as

$$\mathbf{p}(\mathbf{x}) \triangleq \left[\ \tan^{-1}\left(\frac{y-y_0}{x-x_0}\right),\ \ldots,\ \tan^{-1}\left(\frac{y-y_{N-1}}{x-x_{N-1}}\right)\ \right] \tag{10.3}$$

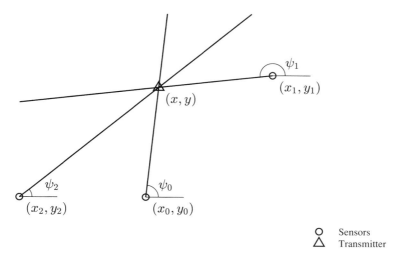

Figure 10.2 Geometry for triangulation with three DF sensors.

and **n** represents the error terms for each of the N bearing measurements. We assume that the measurements are approximately Gaussian, and are unbiased, such that **n** is distributed

$$\mathbf{n} \sim \mathcal{N}\left(\mathbf{0}_N, \mathbf{C}_\psi\right) \tag{10.4}$$

In the general case, $\mathbf{C}_\psi$ can be any covariance matrix, but we often assume that the errors in each bearing measurement are uncorrelated, in which case $\mathbf{C}_\psi$ is diagonal.[2]

Under these assumptions, we can write the log likelihood function of a set of bearing measurements $\boldsymbol{\psi}$ given the actual source location **x** (ignoring scalars and terms that do not vary with **x** or $\boldsymbol{\psi}$).

$$\ell(\boldsymbol{\psi}|\mathbf{x}) = -\frac{1}{2}(\boldsymbol{\psi} - \mathbf{p}(\mathbf{x}))^H \mathbf{C}_\psi^{-1} (\boldsymbol{\psi} - \mathbf{p}(\mathbf{x})) \tag{10.5}$$

The log-likelihood can be evaluated with the function triang.loglikelihood in the provided MATLAB® package, and a noise-free measurement can be quickly generated with the function triang.measurement.

---

[2] Recall from Chapters 7 and 8 that the error terms are approximately Gaussian if they are sufficiently small, but rapidly break down when they exceed a few degrees.

The solutions below will rely on the optimization of either this log likelihood function (in the case of the maximum likelihood) or some other cost metric, such as the distance between the solution and each of the estimated lines of bearing.

10.2.1 3-D Geometry

This formulation is restricted to 2-D geometry (x, y). Extension to 3-D is straightforward, although it can be complex, and there are many possible conventions depending on how the AOAs are reported, such as with cone and rotation angles relative to each system's nominal pointing angle (as discussed in Chapter 8). The simplest extension of the formulation presented here is to represent all of the AOA measurements as a bearing (relative to some nominal angle, such as counterclockwise from north or clockwise from east), and an elevation or depression angle ϕ relative to that (x, y) plane. The altitude coordinate z can thus be represented as

$$\phi_i = \tan^{-1}\left(\frac{z - z_i}{\sqrt{(x - x_i)^2 + (y - y_i)^2}}\right) \quad (10.6)$$

A principal benefit of working in three dimensions is the ability to natively represent the sensors, lines of bearing, and estimated source position in geographic coordinates (such as degrees latitude, degrees longitude, and elevation). Native use of geographic coordinate systems is complicated by their dependence on where in the world the systems are location (a degree of longitude is not the same at every latitude), and calculations of distance or bearing angle are needlessly complex. The notation here (x, y, z) is most closely matched to *Earth-centered Earth-fixed* (ECEF) coordinates, which is a Euclidean 3-D geometry with its origin at the center of the Earth, and all three dimensions expressed in meters. Distances and angles in ECEF are the same as formulated in this section.

The first step is often to take the relative (x, y, z) or (ψ, ϕ) coordinates from each sensor and translate them to the same ECEF reference frame. When a solution is derived, it can then be translated to a more convenient form for expression, such as latitude, longitude and altitude (LLA) or east-north-up (ENU). Interested readers are referred to Section 2.3 of [4] or to the online text [6] for an overview of the different coordinate systems and how to translate between them.

10.3 SOLUTION

In this section, we present several solutions to inverting (10.2). The first two are based on a naive solution from geometric principles. Then, we derive the maximum likelihood solution and present two iterative approximations.

10.3.1 Geometric Solution for Two Measurements

With two AOA solutions, the estimated source position $\hat{\mathbf{x}}$ is given by their intersection. An algebraic solution can be found by solving for the equation that defines the two lines in $y = mx + b$ form, and then solving the system of two equations:

$$\begin{bmatrix} y = \frac{\sin(\psi_0)}{\cos(\psi_0)} x + \left(y_0 - \frac{\sin(\psi_0)}{\cos(\psi_0)} x_0 \right) \\ y = \frac{\sin(\psi_1)}{\cos(\psi_1)} x + \left(y_1 - \frac{\sin(\psi_1)}{\cos(\psi_1)} x_1 \right) \end{bmatrix} \quad (10.7)$$

This solution is implemented in the enclosed MATLAB® code, under the function `utils.find_intersect`.

10.3.2 Geometric Solutions for Three or More Measurements

When three or more noisy AOA estimates are provided, it is virtually impossible for them to intersect in a point, but each set of three AOA estimates will form a triangle. It is straighforward, then, to compute one of the known *centers* of a given triangle, and declare that point the estimated source position. Figure 10.3 plots two of the most well-known triangle centers, the *centroid* (computed by drawing a line from each vertex to the opposite median, and then solving for their intersection) and the *incenter* (found by bisecting each of the three vertices and solving for their intersection).

MATLAB® code is provided for both the centroid and the angle bisector approach in the functions `triang.centroid` and `triang.angle_bisector`. The output of a simple test case, with three sensors, is plotted in Figure 10.4.

10.3.3 Maximum Likelihood Estimate

To formulate the ML estimate, we recall the log-likelihood equation of $\boldsymbol{\psi}$

$$\ell(\boldsymbol{\psi}|\mathbf{x}) = -\frac{1}{2} (\boldsymbol{\psi} - \mathbf{p}(\mathbf{x}))^T \mathbf{C}_{\boldsymbol{\psi}}^{-1} (\boldsymbol{\psi} - \mathbf{p}(\mathbf{x})) \quad (10.8)$$

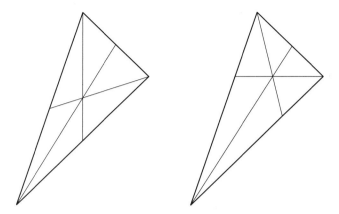

Figure 10.3 Illustration of the (a) centroid and (b) incenter centers of a triangle.

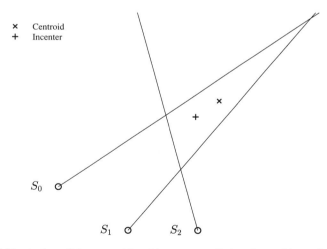

Figure 10.4 Illustration of the centroid and incenter applied to the problem of triangulation with three AOA sensors.

and take the derivative with respect to the target position x

$$\nabla_{\mathbf{x}}\ell(\boldsymbol{\psi}|\mathbf{x}) = \mathbf{J}(\mathbf{x})\mathbf{C}_{\psi}^{-1}\left(\boldsymbol{\psi} - \mathbf{p}(\mathbf{x})\right) \qquad (10.9)$$

where $\mathbf{J}(\mathbf{x})$ is the Jacobian of the system equation $\mathbf{p}(\mathbf{x})$, given as[3]

$$\mathbf{J}(\mathbf{x}) \triangleq \begin{bmatrix} \nabla_{\mathbf{x}}p_0(\mathbf{x}), & \nabla_{\mathbf{x}}p_1(\mathbf{x}), & \ldots, & \nabla_{\mathbf{x}}p_{N-1}(\mathbf{x}) \end{bmatrix} \qquad (10.10)$$

$$\nabla_{\mathbf{x}}p_i(\mathbf{x}) = \frac{1}{\|\mathbf{x} - \mathbf{x}_i\|^2}\begin{bmatrix} -(y - y_i), & x - x_i \end{bmatrix}^T \qquad (10.11)$$

The Jacobian can be called directly with the function triang.jacobian in the provided MATLAB® package, and the ML estimate can be estimated (via a brute force search) with triang.mlSoln.

Recall, the ML estimate is found by setting $\nabla_{\mathbf{x}}\ell(\psi|\mathbf{x})$ equal to zero, and solving for $\mathbf{x}$. Unfortunately, because $\mathbf{J}(\mathbf{x})$ is not linear in $\mathbf{x}$, there is no analytic solution. We discuss two iterative solutions in the following sections.

10.3.4 Iterative Least Squares

The most straightforward approach to solving (10.8) is to posit an estimated position, and then use a Taylor series approximation to update it. If the estimated position is close enough to the true position, then the problem is locally convex, and an iterative solution will eventually converge to the true position [7]. Since we desire a linear solution, we will use the first-order Taylor series approximation about the position estimate $\mathbf{x}$, in vector form. For some function $f(\mathbf{x})$, this is written as

$$f(\mathbf{x} + \Delta\mathbf{x}) \approx f(\mathbf{x}) + \Delta\mathbf{x}^T \nabla_{\mathbf{x}} f(\mathbf{x}) \qquad (10.12)$$

We apply this approximation to the AOA vector $\mathbf{p}(\mathbf{x})$, whose gradient is stored in the Jacobian matrix $\mathbf{J}(\mathbf{x})$. Using this, we can express the linear approximation to (10.3) about some position estimate $\mathbf{x}^{(i)}$, and with offset $\Delta\mathbf{x}^{(i)}$ defined as the difference between $\mathbf{x}$ and the current estimate $\mathbf{x}^{(i)}$

$$\Delta\mathbf{x}^{(i)} = \mathbf{x} - \mathbf{x}^{(i)} \qquad (10.13)$$

$$\mathbf{p}\left(\mathbf{x}^{(i)} + \Delta\mathbf{x}^{(i)}\right) = \mathbf{p}\left(\mathbf{x}^{(i)}\right) + \mathbf{J}^T\left(\mathbf{x}^{(i)}\right)\Delta\mathbf{x}^{(i)} \qquad (10.14)$$

[3] The gradient solution relies on the fact that $\partial \tan^{-1}(x)/\partial x = 1/(1+x^2)$.

We define the measurement offset vector as the difference between the expected AOA vector from the point $\mathbf{x}^{(i)}$ and the (noisy) measured AOA vector $\mathbf{\psi}$.

$$\mathbf{y}\left(\mathbf{x}^{(i)}\right) = \mathbf{\psi} - \mathbf{p}\left(\mathbf{x}^{(i)} + \Delta\mathbf{x}^{(i)}\right) \qquad (10.15)$$

This leads to the linear equation

$$\mathbf{y}\left(\mathbf{x}^{(i)}\right) = \mathbf{J}^T\left(\mathbf{x}^{(i)}\right)\Delta\mathbf{x}^{(i)} + \mathbf{n} \qquad (10.16)$$

where $\mathbf{n}$ is the same angle measurement noise vector from (10.2). In this context, $\mathbf{y}\left(\mathbf{x}^{(i)}\right)$ represents the residual error between the AOA measurements predicted from the current position estimate, and that which was actually observed. The solution to this linear equation will yield an estimate of the offset $\Delta\mathbf{x}$ between the current estimate and the true position of the emitter.

Before we invert (10.16), we whiten the error term $\mathbf{n}$.[4] This is done by premultiplying $\mathbf{y}\left(\mathbf{x}^{(i)}\right)$ with $\mathbf{C}_\psi^{-1/2}$, where $\mathbf{C}_\psi^{-1/2}$ is defined as the matrix that satisfies the equation

$$\mathbf{C}_\psi^{-1/2}\mathbf{C}_\psi\mathbf{C}_\psi^{-1/2} = \mathbf{I}_N \qquad (10.17)$$

This leaves the whitened residual

$$\widetilde{\mathbf{y}}\left(\mathbf{x}^{(i)}\right) = \mathbf{C}_\psi^{-1/2}\left[\mathbf{\psi} - \mathbf{p}\left(\mathbf{x}^{(i)}\right)\right] \qquad (10.18)$$

$$= \mathbf{C}_\psi^{-1/2}\mathbf{J}^T\left(\mathbf{x}^{(i)}\right)\Delta_\mathbf{x}^{(i)} + \widetilde{\mathbf{n}} \qquad (10.19)$$

To solve this, we apply the well-known least squares solution [7]

$$\Delta\mathbf{x}^{(i)} = \left[\mathbf{J}\left(\mathbf{x}^{(i)}\right)\mathbf{C}_\psi^{-1}\mathbf{J}^T\left(\mathbf{x}^{(i)}\right)\right]^{-1}\mathbf{J}\left(\mathbf{x}^{(i)}\right)\mathbf{C}_\psi^{-1}\mathbf{y}\left(\mathbf{x}^{(i)}\right) \qquad (10.20)$$

This result was proposed for AOA triangulation by Torrieri [8, 9], with a slightly altered form of the Jacobian $\mathbf{J}$.

In order to guarantee that this solution is not underdefined, the Jacobian matrix $\mathbf{J}$ must not have more columns than rows. In other words, the number

---

4 Whitening the noise term is a common technique in signal processing, particularly in detection and estimation theory, in that it is the solution to maximization of the SNR in the presence of correlated or nonuniform noise. In this case, it is being used to discount the errors on angle estimates with a higher error variance, and to emphasize errors on angle estimates that have a lower variance.

of AOA measurements must be greater than or equal to the number of spatial dimensions (in this case, 2).

The least square solution in (10.20) is not guaranteed to be optimal, only to be closer than the initial estimate $\left(\mathbf{x}^{(i)}\right)$ under the assumption that $\mathbf{x}^{(i)}$ is sufficiently close to the true position that the problem is indeed locally convex. In order to obtain an accurate solution, the process must be repeated in an iterative fashion. An implementation of this algorithm is included in the provided MATLAB® code, under the function `triang.lsSoln`.

The principal limitations of this approach are the need for repeated calculations to reach convergence and the need for a sufficiently accurate initial position estimate. It is very important to understand that "sufficiently accurate" is ill defined, and there is thus little guarantee that an initial position estimate used to seed this approach has satisfied the requirement for local convexity.

Furthermore, this technique suffers from robustness concerns, namely that it is possible for the estimate to diverge wildly from the true emitter location. Constraints and other optimizations are often necessary to improve accuracy.

10.3.5 Gradient Descent

Gradient descent algorithms compute the direction of steepest gradient at each step and use that to determine the direction in which to adjust the estimated solution. They are a convenient formulation for convex applications and provide faster convergence than iterative least squares [7].

The general descent solution is to update $\mathbf{x}$ with the equation

$$\mathbf{x}^{(i+1)} = \mathbf{x}^{(i)} + t\Delta\mathbf{x}^{(i)} \tag{10.21}$$

where t is a suitably chosen step size, and $\Delta\mathbf{x}^{(i)}$ is the descent direction. In *gradient descent* algorithms, the descent direction is chosen as the negative of the gradient $\left(\Delta\mathbf{x}^{(i)} = -\nabla_\mathbf{x} f(\mathbf{x}^{(i)})\right)$ for some objective function $f(\mathbf{x})$. We define the objective function $f(\mathbf{x})$ to minimize as the norm of the whitened residual $\tilde{\mathbf{y}}(\mathbf{x})$ from (10.18) as

$$f(\mathbf{x}) = \|\tilde{\mathbf{y}}(\mathbf{x})\|_2^2 = (\boldsymbol{\psi} - \mathbf{p}(\mathbf{x}))^T \mathbf{C}_{\boldsymbol{\psi}}^{-1} (\boldsymbol{\psi} - \mathbf{p}(\mathbf{x})) \tag{10.22}$$

This is equivalent to the negative of the log-likelihood function $\ell(\boldsymbol{\psi}|\mathbf{x})$. Thus, the gradient is given as

$$\nabla_\mathbf{x} f(\mathbf{x}) = -\nabla_\mathbf{x} \ell(\boldsymbol{\psi}|\mathbf{x}) = -\mathbf{J}(\mathbf{x})\mathbf{C}_{\boldsymbol{\psi}}^{-1}(\boldsymbol{\psi} - \mathbf{p}(\mathbf{x})) \tag{10.23}$$

The step size can be chosen from among several approaches discussed in [7]. *Exact line search* involves an optimization within each iteration of the gradient descent algorithm, where the step size t is chosen as the point along the gradient $\Delta \mathbf{x}$ that minimizes the objective function $f(\mathbf{x})$

$$t = \arg \min_{s \geq 0} f(\mathbf{x} + s\Delta \mathbf{x}). \tag{10.24}$$

Code for a gradient descent solution to triangulation, which relies on a variant of line search called backtracking line search, is provided in the MATLAB® package, under the function triang.gdSoln.

The function triang.gdSoln accepts parameters alpha and beta, for backtracking line search in the gradient descent update step.[5]

The gradient descent algorithm provided makes one modification to improve numerical stability. The descent direction is defined to have unit norm

$$\Delta \mathbf{x} = -\frac{\nabla_{\mathbf{x}} f(\mathbf{x})}{|\nabla_{\mathbf{x}} f(\mathbf{x})|} \tag{10.25}$$

Backtracking line search can be accessed directly via the provided utility

```
> t = utils.backtrackingLineSearch(f,x_prev,grad,delta_x,alpha,beta);
```

where f is a function handle to the objective function being minimized, x_prev is the starting position, grad is the gradient evaluated at x_prev, and delta_x is the desired direction of travel. alpha and beta are the aforementioned line search coefficients.[6]

Perhaps the most important feature of gradient descent algorithms is that the update equation (10.21) does not contain a matrix inversion that must be updated at each iteration,[7] so it is much faster to compute than the iterative least square solution (10.20).

The principal downside is still the need for a suitably accurate initial estimate, such that convexity of the objective function can be assumed.

Example 10.1 AOA Triangulation

Consider three direction-finding receivers spaced in a line 30 km apart, and a source located 45 km down range and half-way between the two receivers. Find

---

[5] According to [7], good values for alpha range between .01 and .3, while for beta they range between .5 and .8.

[6] In the provided MATLAB® code, we modify the backtracking line search from [7] to ensure that the starting condition for the line search is sufficiently large.

[7] The inverse $\mathbf{C}_{\psi}^{-1}$ can be computed once, before the iterative process.

the estimated source position using each of the four techniques above, given that the AOA measurements are independent and have a standard deviation of 5 degrees ($\sigma_\psi = 5\pi/180 \approx 87$ mrad).[8]

Use a Monte Carlo trial to compute the expected RMSE of each technique. Which approaches the CRLB most quickly?

To begin, we establish the source and sensor positions, and generate 1,000 random samples of the AOA vector ψ.

```
x_source = [15;45]*1e3;              % source position
x_sensor = [-30, 0, 30; 0, 0, 0]*1e3; % sensor positions
C_psi = (5*pi/180)^2*eye(3);         % covariance matrix
num_MC = 1000;
p = triang.measurement(x_sensor,x_source);
psi = p + C_psi.^(1/2)*randn(3,num_MC); % noisy measurements
```

Next, we step through each of the Monte Carlo trials and use the noisy measurements to compute triangulation solutions using each of the four methods; then we store the actual errors and compute the RMSE.

```
for ii=1:num_MC
    this_psi = psi(:,ii);

    % Compute solutions using each of the four methods
    x_ls = triang.lsSoln(x0,psi(:,ii),C_psi,x_init);
    x_grad = triang.gdSoln(x0,psi(:,ii),C_psi,x_init);
    x_ctr = triang.centroid(x,psi(:,ii));
    x_inc = triang.angle_bisector(x,psi(:,ii));

    % Compute error for each solution
    err_ls(:,ii) = x_src - x_ls;
    err_grad(:,ii) = x_src - x_grad;
    err_ctr(:,ii) = x_src - x_ctr;
    err_inc(:,ii) = x_src - x_inc;
end
```

Then, we compute the covariance matrix and use utils.computeCEP50 to compute the CEP50 for each of the solutions.

Figure 10.5(a) plots the geometry of this example, along with the 1-σ confidence interval for each of the three sensors' AOA estimate, and the expected 90% error ellipse centered on the true position (based on the CRLB, which we derive in

---

[8] 1 *mrad* refers to 1 milliradian, or .001 radians, which translates to approximately 1/20 of a degree.

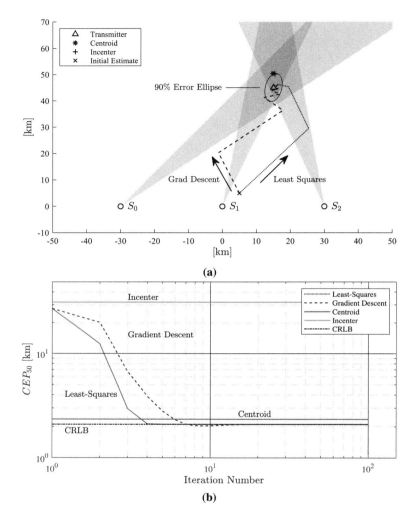

Figure 10.5 Graphical illustration and error as a function of the number of iterations for Example 10.1: (a) Laydown of the three AOA sensors, their noisy estimates, and both geometric and iterative solutions. The 90% error ellipse (based on the CRLB) and the true source position are also plotted, for reference. (b) CEP_{50} for the incenter, centroid, least squares, and gradient descent are plotted for 1,000 Monte Carlo trials against the CEP_{50} from the CRLB. The incenter and centroid are noniterative, so their errors are constant. For the least squares and gradient descent solutions, we actually plot CEP_{50} plus the square of the bias, since early iterations are closely clustered about the seed position $\mathbf{x}^{(0)}$.

Section 10.4). In addition, we plot an example of the estimated solutions for one of the Monte Carlo trials. For the iterative solutions, the first 100 steps are plotted as a dashed line indicating the progression of estimates as the algorithm is iterated.

Figure 10.5(b) plots the CEP_{50} computed across 1,000 Monte Carlo trials for each of the methods discussed, along with the CRLB. For this example, the centroid solution vastly outperforms the incenter, but performs slightly worse than the iterative methods, both of which come very close to achieving the CRLB after just four iterations for the iterative least square solution and 40 iterations for the gradient descent. Between the two iterative methods, the least squares algorithm begins to approach more quickly and is slightly more accurate. The dip in CEP_{50} for the gradient descent algorithm is an artifact of the geometry, and the fact that it passes by the true solution on the way to the convergence point.

10.4 OTHER SOLUTIONS

There are many solutions available in addition to the ones presented here, including those based on minimizing the distance between each of the LOBs and the estimated source position, those based on a grid-based calculation of the probability of the a source occupying various coordinates, and additional techniques for iterative solutions (both additional formulations and alternate numerical algorithms). See Chapter 2 of [3] for a more thorough treatment of triangulation algorithms, and [7] for general discussion of numerical techniques.

10.5 PERFORMANCE ANALYSIS

To analyze performance, we look again to the CRLB. We begin with the FIM from (6.53), and note that the covariance matrix $\mathbf{C}_\psi$ is assumed to be independent of the target position $\mathbf{x}$, so we can ignore the first term.

$$\mathbf{F}_\mathbf{x}(\psi) = \mathbf{J}(\mathbf{x})\mathbf{C}_\psi^{-1}\mathbf{J}^T(\mathbf{x}) \tag{10.26}$$

where the Jacobian $\mathbf{J}(\mathbf{x})$ is given in (10.10), and the CRLB states that the error covariance is lower-bounded by the inverse of $\mathbf{F}_\mathbf{x}(\psi)$

$$\mathbf{C}_\mathbf{x} \geq \left[\mathbf{J}(\mathbf{x})\mathbf{C}_\psi^{-1}\mathbf{J}^T(\mathbf{x})\right]^{-1} \tag{10.27}$$

Example 10.2 Two-Sensor DF System Performance

Consider a two-sensor direction-finding system with a 20-km baseline, and angle measurements with a 2.5-degree standard deviation, representative of many of the legacy DF systems discussed in Chapter 7. What is the maximum cross-range offset x such that a source $y = 100$ km down-range will have $\text{CEP}_{50} \leq 25$ km. To solve this, we compute the CRLB at a series of points with down-range positions of 100 km. First, we define the sensor and source positions.

```
x0 = [-10, 10; 0, 0];                    % Define sensor positions [km]
x_src = [-100:100; 100*ones(1,201)];     % Define source positions [km]
```

The next step is to define the sensor errors and call the CRLB. Finally, for each CRLB estimate, we compute the CEP_{50} using the utility defined in Chapter 9.

```
sigma_psi = 2.5*pi/180;      % 2.5 degrees standard deviation
C_psi = sigma_psi^2*eye(2);  % Error covariance matrix

% Compute the CRLB [m^2] and CEP50 [m]
CRLB = triang.crlb(x0*1e3,x_src*1e3,C_psi);
CEP50 = reshape(utils.computeCEP50(CRLB),size(x_src));
```

At this point, we simply search over the computed CEP_{50} values, find all those that match the desired performance (< 25 km), and compute the maximum off-axis source position

```
good_points = CEP50 <= 25e3;      % binary mask
max_cross_range = max(abs(x_src(1,good_points)));
```

This results in a value of 35 km. If the down-range position is 100 km, then any sources with a cross-range position less than 35 km will meet the desired performance ($\text{CEP}_{50} < 25$ km). For illustration purposes, we plot the CEP_{50} at these points and other down range positions (between -100 km and 100 km) as a series of contours in Figure 10.6. As we see in Figure 10.6, the shape of the 25-km contour (and the others) varies greatly with down range position. At the 100-km standoff specified, it is expanding as down range reduces, to a maximum of almost 70-km cross-range when the down range position is roughly 50 km. But, for ranges closer than 50 km, it rapidly collapses. The reason for this is that all source positions where the cross-range is much larger than the down range will have very similar AOA estimates from the two sources. This leads to a geometry where the two estimates have little information on the distance of the target. This effect is referred to as *geometric dilution of precision* [3, 8].

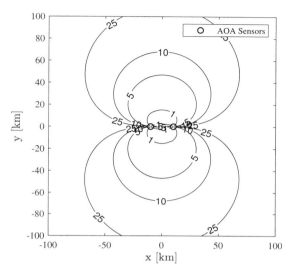

Figure 10.6 CEP$_{50}$ based on the CRLB of a two-sensor DF triangulation system, from Example 10.2.

Example 10.3 Multibaseline AOA Performance

Repeat the analysis from Example 10.2, but in this case, consider that a third DF station is 10-km downrange and halfway between the two existing sensors. In this situation, how close to the origin of the DF system must a source be, to support CEP$_{50}$ ≤ 25 km, regardless of the direction?

We begin with the same approach as before, defining the sensor positions, and a set of possible source positions. In this case, we'll use 500-m spacing in both dimensions for the potential source positions, and compute the CRLB and CEP$_{50}$ for each point

```
x0 = [-10, 10; 0, 0];           % Sensor positions [km]
src_vec = -100:.5:100;          % Source position sweep [km]
[X_src,Y_src] = ndgrid(src_vec); % Interleave sweep for x and y dims
x_src = [X_src(:),Y_src(:)]';   % Source positions [km]

% Compute CRLB [m^2] and CEP50 [m]
CRLB = triang.CRLB(xs*1e3,x_src*1e3,C_psi);
CEP50 = utils.computeCEP50(CRLB);
```

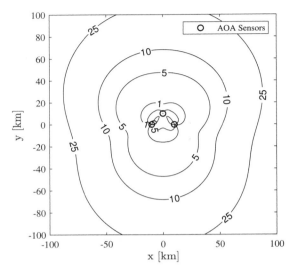

Figure 10.7 CEP$_{50}$ based on the CRLB of a three-sensor DF triangulation system, from Example 10.3.

Once again, we find the mask of all points that satisfy the criteria, and compute the distance from the origin to each test point.

```
good_points = CEP50 <= 25e3;            % binary mask
rng = sqrt(sum(abs(x_src).^2,1));       % range from origin
```

Finally, we note that the desired answer (the maximum range at which CEP$_{50}$ $\leq$ 25 km, regardless of AOA) is given by the minimum range at which at least one angle does not satisfy it. Thus, we compute

```
max_range = min(rng(~good_points));     % max range when mask is false
```

which yields an answer of 80.6 km.

10.6 PROBLEM SET

10.1 Using the provided MATLAB® code, generate the measurements for a three-sensor triangulation system with receivers at $x_0 = [3, 5]$, $x_1 = [2, 6]$, $x_2 = [0, 0]$, and a source at $x_s = [4, 7]$. Add noise with standard deviation $\sigma_\psi = 40$

mrad to each measurement. Use the centroid and bisector methods to generate an estimate of the target position.

10.2 Generate 100 unique random measurements for the same scenario as in the last problem. Run the centroid and bisector methods on each and compute the RMSE for each approach.

10.3 For the same set of 100 random measurements as in the last problem, compute the ML estimate with a brute force search between the origin $[0, 0]$ and the point $[10, 10]$, with desired grid spacing of 0.2, using the provided MATLAB® function triang.mlSoln. Compute the CEP_{50}.

10.4 Compute the CRLB for the measurement system in Problem 10.1.

10.5 For the same measurement system as before, compute the iterative least squares and gradient descent estimates, with a maximum of 10,000 iterations. Plot the full set of intermediate estimates for each, and overlay with the true target position, and 90% error ellipse drawn from the CRLB.

10.6 Consider a two-sensor direction finding system with a 10-km baseline, and angle measurements with a 1.5-degree standard deviation. What is the maximum cross-range offset x such that a source at $y = 25$ km down range will have $CEP_{50} \leq 10$ km.

10.7 Repeat the previous problem with a 2.5-degree standard deviation on the error measurements.

References

[1] D. Adamy, *EW 103: Tactical battlefield communications electronic warfare.* Norwood, MA: Artech House, 2008.

[2] A. Graham, *Communications, Radar and Electronic Warfare.* Hoboken, NJ: John Wiley & Sons, 2011.

[3] R. Poisel, *Electronic warfare target location methods.* Norwood, MA: Artech House, 2012.

[4] F. Guo, Y. Fan, Y. Zhou, C. Xhou, and Q. Li, *Space electronic reconnaissance: localization theories and methods.* Hoboken, NJ: John Wiley & Sons, 2014.

[5] R. Stansfield, "Statistical theory of df fixing," *Journal of the Institution of Electrical Engineers-Part IIIA: Radiocommunication*, vol. 94, no. 15, pp. 762–770, 1947.

[6] M. Gimond. Intro to GIS and Spatial Analysis. [Online]. Available: https://mgimond.githaub.io/Spatial/index.html

[7] S. Boyd and L. Vandenberghe, *Convex optimization.* Cambridge, UK: Cambridge University Press, 2004.

[8] D. J. Torrieri, "Statistical theory of passive location systems," *IEEE Transactions on Aerospace and Electronic Systems*, vol. AES-20, no. 2, pp. 183–198, March 1984.

[9] D. Torrieri, "Statistical theory of passive location systems," Army Materiel Development and Readiness Command, Countermeasures/Counter-Countermeasures Office, Tech. Rep. ADA128346, 1983.

Chapter 11

TDOA

TDOA is a high-precision emitter location technique that is frequently used to generate tracks of noncooperative emitting targets over time. Due to its very high precision, and completely passive nature, TDOA is a cornerstone of modern air defense systems.

The basic premise of TDOA is to measure the difference in time of arrival of a pulse at several sensor locations, and to use that measurement to compute the emitter's position. Most algorithms solve the pairwise hyperbolas that trace constant time difference of arrival, and look for the intersection of the full set of computed hyperbolas, although other formulations exist and are discussed. For this reason, these techniques fall into a class of geolocation algorithms sometimes referred to as *hyperbolic geolocation algorithms.*

This chapter discusses the motivations and geometric formulation of TDOA geolocation, presents several algorithms to solve the TDOA system of equations, and derives geolocation error performance.

11.1 BACKGROUND

Interest in passive geolocation techniques is as old as the field of electronic warfare. Early solutions to TDOA emitter location are traced as far back as the 1940s [1–5]; but its first documented use was in World War I, where microphones were used to record gunshots and trace the position of hidden artillery via TDOA methods [1]. TDOA, and other hyperbolic position location schemes, have the potential to provide very high accuracy at very long ranges. With time of arrival accuracy on the order of 1 μs, a TDOA system with a 10-km baseline[1] could provide position accuracy on the order of 50 km out to a range of 300 km, sufficient for early warning or to cue more accurate sensor systems. Achieving such high-accuracy TDOA measurements is not straightforward, but this example illustrates the tremendous potential of hyperbolic position location systems, especially when one considers the completely passive nature of TDOA, and consequently the relative cost of a TDOA constellation when compared with a high-powered radar system.

The formulation for TDOA processing is essentially the same as for GPS position estimation; both rely on the fact that electromagnetic waves travel at the speed of light. The difference is primarily that the transmitter and receivers are reversed: in GPS there are multiple emitters, and it is the sensor that is attempting to estimate its own position. The principal advantage of GPS is that the emitters are transmitting a known signal and a reference clock that can be used for time of arrival calculation, while TDOA systems are typically employed against noncooperative systems that are using unknown waveforms and emitting at unknown times.

11.2 FORMULATION

Signals propagate according to some velocity (for electromagnetic waves, this is the speed of light), and arrive at the receiving stations with some time delay, proportional to the range between the emitter and receiver. If the signal transmitted is $s(t)$, then the one received at the ith sensor is defined as

$$y_i(t) = \alpha_i s(t - \tau_i) + n_i(t) \tag{11.1}$$

where (for electromagnetic waves) the delay τ_i is defined $\tau_i = R_i/c$, and R_i is the distance between the emitter and the ith sensor. It is very straightforward to solve for R_i. The source's location is computed by finding the intersection across all sensors of all points that satisfy the measured distance to that sensor, as shown in Figure 11.1(a). The principal limitation of this approach is that it requires knowledge of when the signal was transmitted, a piece of information that is rarely available in EW.

In order to compensate for this lack of information, one can use time TDOA calculations. For each pair of sensors, the set of points that satisfy the measured TDOA is a hyperbola, often referred to as an isochrone. Just as with the time of arrival solution, the sensor's location is computed by finding the intersection of each of the TDOA hyperbolas, as shown in Figure 11.1(b).

1   The baseline of a system is the maximum separation between receivers.

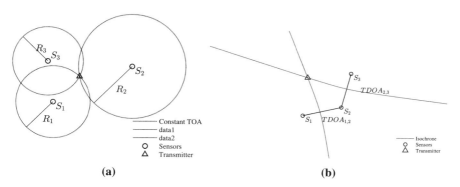

Figure 11.1 Geometry for determination of emitter location based on (a) time of arrival, and (b) TDOA

11.2.1 Isochrones

Consider any two sensors, S_m and S_n. We define the time difference measurement between them as $t_{m,n} = \tau_m - \tau_n$. As mentioned in Section 11.2, it is straightforward to make the transformation from time difference to range difference: $R_{m,n} = ct_{m,n}$. The time difference between two signals sketches out a hyperbola defined by the equation

$$R_{m,n}(x,y) = \sqrt{(x - x_m)^2 + (y - y_m)^2} - \sqrt{(x - x_n)^2 + (y - y_n)^2} \quad (11.2)$$

Figure 11.2 shows a trace of several TDOA hyperbolas, called isochrones, between two sensors. Each isochrone represents a constant TDOA between the two sensors.

11.2.2 Number of Sensors

The time difference of any two sensors can trace out a hyperbola in 2-D space, but a third sensor is required to arrive at a single sensor location, as in Figure 11.1(b). To extend this to three dimensions, a fourth sensor is required. While this is the bare minimum, we will show later in this chapter that an additional sensor is recommended (three total for 2-D and four for 3-D). This is because when the source is aligned such that two of the sensors have the same line of sight, they are ill-posed to perform TDOA, and the rank of the data matrix is reduced by one. Adding an extra sensor, and ensuring that no three sensors fall on a line, prevents this situation.

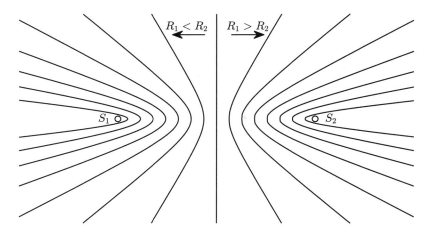

Figure 11.2 Illustration of isochrones for a pair of sensors. Contours are lines of constant time difference of arrival.

Furthermore, many of the proposed algorithms require a second additional sensor to simplify the system of equations.

11.3 SOLUTION

The isochrone equation in (11.2) is distinctly nonlinear. This will present a challenge for TDOA geolocation algorithms, and we will see that there is indeed no closed-form solution. Nevertheless, several algorithms have been proposed to solve for the emitter location, with varying degrees of accuracy and complexity. Before we define any of the algorithms, we first present a common framework for the problem.

For simplicity, we rewrite (11.2) in vector form, using the ℓ_2 (Euclidean) norm

$$R_{m,n}(\mathbf{x}) = \|\mathbf{x} - \mathbf{x}_m\|_2 - \|\mathbf{x} - \mathbf{x}_n\|_2 \tag{11.3}$$

where the $\mathbf{x}$ is the emitter location (in either two- or three-dimensional coordinates), and $\mathbf{x}_m$ is the location of the mth sensor. $R_{m,n}(\mathbf{x})$ is the range difference to a source at location $\mathbf{x}$, as observed by the mth and nth sensors, computed from the comparison of the time of arrival at the two sensors. From this, we construct the range difference vector $\mathbf{r}(\mathbf{x})$, with all ranges referenced to the Nth sensor

$$\mathbf{r}(\mathbf{x}) = [R_{1,N}(\mathbf{x}),\ \ldots,\ R_{N-1,N}(\mathbf{x})]^T \tag{11.4}$$

This vector has $N-1$ entries, each referenced to the Nth sensor. While it is possible to compute time difference measurements for all of the sensor pairs, the measurements are redundant; there are only $N-1$ unique TDOA measurements. For simplicity, we assume that the Nth sensor is to be used as the reference sensor, against which all others are compared.

At this point, we introduce our measurement vector ρ, a noisy version of the true range difference vector $\mathbf{r}$

$$\rho = \mathbf{r}(\mathbf{x}) + \mathbf{n} \tag{11.5}$$

Recall from Chapter 6 that maximum likelihood estimation relies on finding the unknown parameter $\mathbf{x}$ that maximizes the likelihood of observing the data ρ that was measured. Thus, we need to compute the probability density function of ρ. It has been shown that TDOA estimates are approximately Gaussian [6, 7]. Thus, we assume that $\mathbf{n}$ is a zero-mean Gaussian random vector with covariance matrix $\mathbf{C}_\rho$. If the measurements are uncorrelated, then this will be a diagonal matrix: $\mathbf{C}_\rho = \mathrm{diag}\{\sigma_1^2, \ldots, \sigma_{N-1}^2\}$.

This requires that the TDOA measurements be taken from $N/2$ distinct pairs; no sensor may be used for more than one measurement, as that would introduce correlation between some of the TDOA estimates (from their common source). That approach is suboptimal, since it requires far more sensors to arrive at the required number of TDOA measurements ($2N_{\mathrm{dim}}$ vs. $N_{\mathrm{dim}} + 1$). In the limiting case of N sensors, one of which is used as a common reference, the covariance matrix will take the form

$$\mathbf{C}_\rho = c^2 \begin{bmatrix} \sigma_1^2 + \sigma_N^2 & \sigma_N^2 & \cdots & \sigma_N^2 \\ \sigma_N^2 & \sigma_2^2 + \sigma_N^2 & \cdots & \\ \vdots & & \ddots & \vdots \\ \sigma_N^2 & \cdots & & \sigma_{N-1}^2 + \sigma_N^2 \end{bmatrix} \tag{11.6}$$

where σ_i^2 is the variance in the time of arrival measurements for the ith sensor, and c is the speed of light. From this, we can write the probability density function

$$f_{\mathbf{x}}(\rho) = (2\pi)^{-\frac{(N-1)}{2}} |\mathbf{C}_\rho|^{-1/2} e^{-\frac{1}{2}(\rho - \mathbf{r}(\mathbf{x}))^T \mathbf{C}_\rho^{-1}(\rho - \mathbf{r}(\mathbf{x}))} \tag{11.7}$$

and the log-likelihood function

$$\ell(\mathbf{x}|\rho) = -\frac{1}{2}(\rho - \mathbf{r}(\mathbf{x}))^T \mathbf{C}_\rho^{-1} (\rho - \mathbf{r}(\mathbf{x})) \tag{11.8}$$

The log-likelihood can be evaluated with the function tdoa.loglikelihood in the provided MATLAB® package, and a noise-free measurement can be quickly generated with the function tdoa.measurement.

11.3.1 Maximum Likelihood Estimate

The maximum likelihood estimate is found by maximizing the log-likelihood of ρ given the unknown emitter position x. We take the derivative of the log likelihood with respect to the unknown sensor position x, given as

$$\nabla_\mathbf{x} \ell(\mathbf{x}|\boldsymbol{\rho}) = \mathbf{J}(\mathbf{x}) \mathbf{C}_\rho^{-1} (\boldsymbol{\rho} - \mathbf{r}(\mathbf{x})) \tag{11.9}$$

where $\mathbf{J}(\mathbf{x})$ is the Jacobian matrix of $\mathbf{r}(\mathbf{x})$

$$\mathbf{J}(\mathbf{x}) = [\nabla_\mathbf{x} R_{1,N}, \ldots, \nabla_\mathbf{x} R_{N-1,N}] \tag{11.10}$$

and $\nabla_\mathbf{x} R_{i,N}$ is the gradient of the range difference to the emitter between the ith sensor and the reference sensor

$$\nabla_\mathbf{x} R_{i,N} = \frac{\mathbf{x} - \mathbf{x}_i}{|\mathbf{x} - \mathbf{x}_i|_2} - \frac{\mathbf{x} - \mathbf{x}_N}{|\mathbf{x} - \mathbf{x}_N|_2} \tag{11.11}$$

The Jacobian can be directly computed with tdoa.jacobian in the provided MATLAB® package, and the ML estimate can be approximated with a brute force search via tdoa.mlSoln.

The gradient is not linear in x, and the result is that there is not a closed-form solution to this optimization problem, nor is it convex, which would lend itself to convenient and efficient numerical solutions [8]. Nevertheless, a number of algorithms have been proposed, which we summarize next.

11.3.2 Iterative Least Squares Solution

The most straightforward approach to solving (11.8) is to posit an estimated position, and then use a Taylor series approximation to update it. This is the same approach described for iterative least squares in Chapter 10, with the principal difference being that the AOA estimates $\boldsymbol{\psi}$ are replaced by range-difference of arrival estimates $\boldsymbol{\rho}$ and the AoA system vector $\mathbf{f}(\mathbf{x})$ is replaced by the true RDOA $\mathbf{r}(\mathbf{x})$. The application of this approach to TDOA is discussed in [9].

The gradient of each range difference is collected in the Jacobian matrix $\mathbf{J}(\mathbf{x})$ (11.10). Using this, we can express the linear approximation to (11.4) about some position estimate $\mathbf{x}^{(i)}$, and with offset $\Delta\mathbf{x}^{(i)}$:

$$\mathbf{r}\left(\mathbf{x}^{(i)} + \Delta\mathbf{x}^{(i)}\right) = \mathbf{r}\left(\mathbf{x}^{(i)}\right) + \mathbf{J}^T\left(\mathbf{x}^{(i)}\right)\Delta\mathbf{x}^{(i)} \qquad (11.12)$$

We define the measurement offset vector

$$\mathbf{y}\left(\mathbf{x}^{(i)}\right) = \boldsymbol{\rho} - \mathbf{r}\left(\mathbf{x}^{(i)} + \Delta\mathbf{x}^{(i)}\right) \qquad (11.13)$$

This leads to the linear equation

$$\mathbf{y}\left(\mathbf{x}^{(i)}\right) = \mathbf{J}^T\left(\mathbf{x}^{(i)}\right)\Delta\mathbf{x}^{(i)} + \mathbf{n} \qquad (11.14)$$

where $\mathbf{n}$ is the same range different measurement noise vector from (11.5). In this context, $\mathbf{y}\left(\mathbf{x}^{(i)}\right)$ represents the residual error between the range difference measurements predicted from the current position estimate and that which was actually observed.

Recall that $\mathbf{y}\left(\mathbf{x}^{(i)}\right)$ has covariance matrix $\mathbf{C}_\rho$ and that performance is optimized if this error covariance can be whitened by premultiplication with $\mathbf{C}_\rho^{-1/2}$.

$$\tilde{\mathbf{y}}\left(\mathbf{x}^{(i)}\right) = \mathbf{C}_\rho^{-1/2}\mathbf{J}^T\left(\mathbf{x}^{(i)}\right)\Delta\mathbf{x}^{(i)} + \mathbf{C}_\rho^{-1/2}\mathbf{n} \qquad (11.15)$$

The solution to this linear equation will yield an estimate of the offset $\Delta\mathbf{x}$ between the current estimate and the true position of the emitter. To solve this, we apply the well known least squares solution [9]

$$\Delta\mathbf{x}^{(i)} = \left[\mathbf{J}\left(\mathbf{x}^{(i)}\right)\mathbf{C}_\rho^{-1}\mathbf{J}^T\left(\mathbf{x}^{(i)}\right)\right]^{-1}\mathbf{J}\left(\mathbf{x}^{(i)}\right)\mathbf{C}_\rho^{-1}\mathbf{y}\left(\mathbf{x}^{(i)}\right) \qquad (11.16)$$

In order to guarantee that this solution is not underdefined, the Jacobian matrix $\mathbf{J}$ must not have more columns than rows. In other words, the number of unique TDOA pairs ($N - 1$) must be greater than or equal to the number of spatial dimensions.

The least square solution in (11.16) is not guaranteed to be optimal, only to be closer than the initial estimate $\left(\mathbf{x}^{(i)}\right)$ under the assumption that $\mathbf{x}^{(i)}$ is sufficiently close to the true position that the problem is indeed locally convex. In order to obtain an accurate solution, the process must be repeated in an iterative fashion. An implementation of this algorithm is included in the provided MATLAB® code, under tdoa.lsSoln.

The principal limitations of this approach are the need for repeated calculations to reach convergence and the need for a sufficiently accurate initial position estimate. It is very important to understand that "sufficiently accurate" is ill defined, and there is thus little guarantee that an initial position estimate used to seed this approach has satisfied the requirement for local convexity.

Furthermore, this technique suffers from robustness concerns, namely that it is possible for the estimate to diverge wildly from the true emitter location. To improve numerical stability, we recommend supplying the algorithm with only the diagonal elements of $\mathbf{C}_\rho$, rather than the full covariance matrix

```
C_ls = diag(diag(Cr));
```

Constraints and other optimizations are often necessary to improve accuracy. There is a rich field of research developing new least squares–based solutions to the TDOA problem [10, 11]

11.3.3 Gradient Descent Algorithms

Gradient descent algorithms compute the direction of steepest gradient at each step and use that to determine the direction in which to adjust the estimated solution. They are a convenient formulation for convex applications and provide faster convergence than iterative least squares [8]. Gradient descent algorithms can be applied to the linear least square approximation of TDOA geolocation [12], as applied to triangulation in Chapter 10.

Consider the matrix form in (11.14); the general descent solution is to update $\mathbf{x}$ with the equation

$$\mathbf{x}^{(i+1)} = \mathbf{x}^{(i)} + t\Delta\mathbf{x}^{(i)} \qquad (11.17)$$

where t is a suitably chosen step size, and $\Delta\mathbf{x}^{(i)}$ is the descent direction. In *gradient descent* algorithms, the descent direction is chosen as the negative of the gradient $\left(\Delta\mathbf{x}^{(i)} = -\nabla_\mathbf{x} f(\mathbf{x}^{(i)})\right)$. Given the system matrix in (11.14), we define the objective function $f(\mathbf{x})$ to minimize as the norm of the offset $\mathbf{y}(\mathbf{x})$. In this manner, the gradient is computed

$$f(\mathbf{x}) = \|\mathbf{y}(\mathbf{x})\|_2^2 \qquad (11.18)$$
$$\nabla_\mathbf{x} f(\mathbf{x}) = -2\mathbf{J}(\mathbf{x})\mathbf{y}(\mathbf{x}) \qquad (11.19)$$

To improve performance, we replace the residual $\mathbf{y}(\mathbf{x})$ with the whitened residual $\widetilde{\mathbf{y}}(\mathbf{x})$. In this case, $f(\mathbf{x})$ is a scaled version of the negative of the log-likelihood function $\ell(\mathbf{x}|\boldsymbol{\rho})$.

$$f(\mathbf{x}) = \|\widetilde{\mathbf{y}}(\mathbf{x})\|_2^2 = -2\ell(\mathbf{x}|\boldsymbol{\rho}) \qquad (11.20)$$
$$\nabla_{\mathbf{x}} f(\mathbf{x}) = -2\mathbf{J}(\mathbf{x})\mathbf{C}_\rho^{-1}\mathbf{y}(\mathbf{x}) \qquad (11.21)$$

where the gradient contains reference to the original (unwhitened) error residual $\mathbf{y}(\mathbf{x})$. This formulation can be considered an iterative approximation of the maximum likelihood estimate, since at each stage we choose the descent direction based on the gradient of the log-likelihood.

As discussed in Chapter 10, the step size t is constructed via a line search method (such as *exact line search* or *backtracking line search*), and the descent direction is restricted to be unit norm. The gradient descent solution for TDOA can be called with the function tdoa.gdSoln.

Perhaps the most important feature of gradient descent algorithms is that the update equation (11.17) does not contain a matrix inversion, so it is much faster to compute than the iterative least square solution (11.16).

The principal downside is still the need for a suitably accurate initial estimate, such that convexity of the objective function can be assumed.

11.3.4 Chan-Ho Approach

A closed-form solution to TDOA localization was proposed by Chan and Ho in 1994 [9]. The basic approach is to first require an additional sensor; where traditional TDOA can be completed with $N_{dim} + 1$ sensors (three sensors for 2-D, four for 3-D), the Chan and Ho approach requires $N_{dim} + 2$. The reason for the additional sensor is that Chan and Ho define a nuisance parameter, the distance between the reference sensor and the emitter, and use that additional parameter to linearize the system of equations; the addition of an extra parameter means an extra sensor is necessary to solve the new (larger) system of linear equations.

We define the expanded parameter vector as

$$\boldsymbol{\theta} = \begin{bmatrix} \mathbf{x} \\ R_N(\mathbf{x}) \end{bmatrix} \qquad (11.22)$$

it can be shown that the TDOA system of equations can be represented as the linear equation [9]

$$\mathbf{y} = \mathbf{G}\boldsymbol{\theta} + \mathbf{n} \qquad (11.23)$$

where the shifted measurement vector **y** and system matrix **G** are defined as

$$\mathbf{G} = -\begin{bmatrix} \mathbf{x}_1^T - \mathbf{x}_N^T \\ \vdots \\ \mathbf{x}_{N-1}^T - \mathbf{x}_N^T \end{bmatrix} \tag{11.24}$$

$$\mathbf{y} = \frac{1}{2}\begin{bmatrix} R_{1,N}^2 - \|\mathbf{x}_1\| + \|\mathbf{x}_N\| \\ \vdots \\ R_{N-1,N}^2 - \|\mathbf{x}_{N-1}\| + \|\mathbf{x}_N\| \end{bmatrix} \tag{11.25}$$

Unfortunately, the system cannot be solved directly, since the elements of $\boldsymbol{\theta}$ are related by a nonlinear equation. The proposed solution is to ignore that fact and first solve the system of linear equations as if the elements of $\boldsymbol{\theta}$ were unrelated, and then to go back and apply their relationship to the estimated solution. The reason for this difference is that the error covariance matrix of $\overline{\mathbf{n}}$ depends on the range from the emitter to each of the $N-1$ nonreference sensors. In the far-source case, this can be assumed to be a scaled identity matrix, but in the near-source case it must also be estimated.

In either case, the parameter vector is first estimated with

$$\widehat{\boldsymbol{\theta}} = \left[\mathbf{G}^T\mathbf{C}^{-1}\mathbf{G}\right]^{-1}\mathbf{G}^T\mathbf{C}^{-1}\mathbf{y} \tag{11.26}$$

If the source is near, we then use the estimate of $\boldsymbol{\theta}$ to construct the range matrix **B**, and geolocation error covariance matrix $\widehat{\mathbf{C}}$

$$\mathbf{B} = 2\mathrm{diag}\{\mathbf{R}(\widehat{\mathbf{x}})\} \tag{11.27}$$
$$\widehat{\mathbf{C}} = \mathbf{B}\mathbf{C}\mathbf{B}^T \tag{11.28}$$

where $\widehat{\mathbf{x}}$ is the first N_{dim} elements of $\widehat{\boldsymbol{\theta}}$. The new estimated covariance matrix $\widehat{\mathbf{C}}$ can then be used to update $\widehat{\boldsymbol{\theta}}$ by replacing **C** in (11.26) with $\widehat{\mathbf{C}}$.[2]

The second step is to update the parameter estimate, making use of the relationship between its elements. To do this, we redefine the system matrices using an offset from the estimated position $\widehat{\mathbf{x}}$. The new parameter vector $\boldsymbol{\theta}_1$ for the second stage of the algorithm is

$$\boldsymbol{\theta}_1 = \widehat{\boldsymbol{\theta}} - \begin{bmatrix} \mathbf{x}_N \\ 0 \end{bmatrix} \tag{11.29}$$

---

[2] Although not mentioned in the paper, source code released by the authors of [9] suggests completing this adjustment three times, regardless of whether the estimate is in the near field.

and the updated system of equations is written as

$$\mathbf{y}_1 = \mathbf{G}_1 \boldsymbol{\theta}_1 + \mathbf{n}_1 \tag{11.30}$$

where the covariance matrix of $\mathbf{n}_1$ is

$$\mathbf{C}_1 = \mathbf{B}_1 \left(\mathbf{G}^T \widehat{\mathbf{C}}^{-1} \mathbf{G} \right)^{-1} \mathbf{B}_1^T \tag{11.31}$$

and the updated system matrices are

$$[\mathbf{y}_1]_i = [\boldsymbol{\theta}_1]_i^2 \tag{11.32}$$

$$\mathbf{G}_1 = \begin{bmatrix} \mathbf{I}_N \\ \mathbf{0}_{1 \times N} \end{bmatrix} \tag{11.33}$$

$$\mathbf{B}_1 = 2 \mathrm{diag}\{\boldsymbol{\theta}_1\} \tag{11.34}$$

Note that the parameter vector now has the squared distance between the reference sensor and source in each dimension. The solution of this system of equations will give us the square of the offset from the reference; the initial solution $\widehat{\mathbf{x}}$ can be used to determine in which direction the offsets should be applied.

The offsets are solved

$$\mathbf{A} = \mathbf{G}_1^T \mathbf{B}_1^{-1} \mathbf{G}^T \mathbf{C}_1^{-1} \mathbf{G} \mathbf{B}_1^{-1} \tag{11.35}$$

$$\widehat{\boldsymbol{\theta}}_1 = [\mathbf{A}\mathbf{G}_1]^{-1} \mathbf{A} \mathbf{y}_1 \tag{11.36}$$

$$\widehat{\mathbf{x}}_1 = \pm\sqrt{\widehat{\boldsymbol{\theta}}_1} + \mathbf{x}_N \tag{11.37}$$

The appropriate solution to the square root is chosen based on which value provides an estimate $\widehat{\mathbf{x}}_1$ that is in the same quadrant as the initial position estimate $\widehat{\mathbf{x}}$.

This algorithm is included in the provided MATLAB® code, under the function tdoa.chanHoSoln. Improvements on this algorithm have been proposed, to reduce the bias in the result and improve robustness in special cases; interested readers are referred to [13].

11.3.5 Spherical Methods

There is a class of techniques that seek a different approach to solving the TDOA geolocation problem, referred to a *spherical geolocation* approaches, in contrast to the *hyperbolic geolocation* formulation introduced earlier in this chapter. The principal difference is that while the TDOA at any two receivers defines a hyperbola, the TDOA at any three receivers defines a straight line of position, one that is the major axis of a sphere that intersects the three sensors and has the true emitter position as one of its two focii.

There are two well known approaches based on this geometry, spherical-intersection (S-X) [14], and spherical-integration (SI) [15, 16], although they have been shown to be algebraically equivalent [14]. The basic approach of these algorithms is to rederive the TDOA formulation using squared range difference of arrival, and to perform a matrix projection that has the data vector ρ in its null space, thereby simplifying the system of equations. While the spherical solutions are closed-form, achieving a result much more quickly than the iterative methods shown previously, they do not provide asymptotically efficient solutions—meaning that they do not converge to the theoretical performance bounds at high SNR [9].

For brevity, we do not include these techniques here. Interested readers are referred to [14] and [15].

11.4 TDOA ESTIMATION

Up until this point, we have taken it for granted that there exists a TDOA estimate between each sensor and the reference sensor. If the transmitted signal is known, this is an arbitrary step, as each receiver can utilize a matched filter to obtain a good estimate of the signal's TOA, even at low SNR. For unknown transmit signals, it is much more difficult.

There are two principal approaches to TDOA estimation for unknown sources: (1) *leading edge* or *peak detection* of pulse arrival times (TOA) and comparison of the TOA estimates from all sensors at a central fusion engine, and (2) *cross-correlation* of received signals at a central fusion engine. The former is much simpler to estimate and has a far lower communications overhead, while the latter is much more robust and accurate. We will discuss each technique in turn.

In either case, it has been shown that the error term in TDOA estimates, **n** is approximately Gaussian [6, 7]. The variances are dependent on the forms used below, but the structure of the covariance matrix depends on how the TDOA estimates are computed. In the case of a single reference sensor, as discussed earlier in this chapter, the covariance matrix takes the form of an identity matrix for each of the unique sensor error terms, and a common error term among all estimates from the common reference, defined in (11.6).

Another common scheme is to define unique pairs of TDOA sensors. This has the benefit of providing a diagonal covariance matrix, but requires that the number of sensors be twice the number of dimensions, and does not efficiently use all of the information given. The covariance matrix in this scenario would be given as

$$\mathbf{C} = c^2 \text{diag}\left\{\bar{\sigma}_1^2, \ldots, \bar{\sigma}_N^2\right\} \tag{11.38}$$

where the variances $\bar{\sigma}_n^2$ are the sum of the variances for the two sensors used for measurement n, again translated to range difference by multiplying by the speed of light squared. We now discuss how the time of arrival estimates are obtained.

11.4.1 Time of Arrival Detection

The first approach to TDOA is to independently measure the time of arrival at each sensor, and then transmit only the timestamps and a *pulse descriptor word* to the central processing station for comparison and TDOA processing.[3] If the transmitted signal is known, then the received signal can be run through a matched filter and analyzed for peaks [17]; this is not a likely situation in TDOA. Thus, we must rely on energy-detection schemes. A number of techniques have been proposed in the literature [18–20]; the simplest to illustrate is leading-edge detection with a threshold (see Figure 11.3); however it is very susceptible to noise.

An improved edge detector, based on estimating the shape of the leading edge of the input signal's autocorrelation, is analyzed in [20]. The improved TOA detector has an error variance given by the formula

$$\sigma^2_{\text{TOA,Edge}} = \frac{1}{\xi}\left[1 + 2\frac{\alpha}{\tau} + 2\left(\frac{\alpha}{\tau}\right)^2\right] \tag{11.39}$$

where ξ is the SNR, α is a threshold parameters, and τ is the time delay (in chips) between successive points used to estimate the shape of the leading edge. The other technique studied in [20] is peak detection, based on estimation of the peak of the received signal's autocorrelation function, which results in an error variance given as

$$\sigma^2_{\text{TOA,Peak}} = \frac{1}{\sqrt{2\xi}} \tag{11.40}$$

---

[3] The *pulse descriptor word* is a collection of estimated parameters from the received pulse, and can include features such as the center frequency, bandwidth, modulation scheme, pulse repetition interval, and angle of arrival (if measured). It is the primary method used for data fusion in scenarios where there is more than one emitter within the field of view.

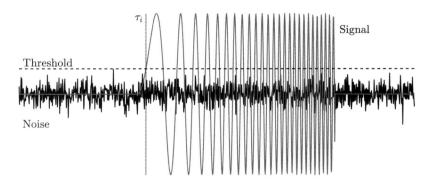

Figure 11.3 Illustration of leading-edge detection for time of arrival estimation.

Peak detection provides a much simpler error estimate, and for most reasonable values of the threshold α and delay τ, it performs better than edge detection. This equation is provided in the function `tdoa.peakDetectionError`. An example is plotted in Figure 11.5.

11.4.2 Cross-Correlation Processing

Cross-correlation TDOA estimation requires that the complex sampled signal be sampled at each sensor, and transmitted to a central fusion engine for processing. At the fusion engine, the received signals for various pairs are correlated against each other, and the peak is located to determine the TDOA between the two sensors, as shown in Figure 11.4.

Performance of the cross-correlation processing is given by the bandwidth of the source signal; signals with a wider bandwidth will exhibit a narrower cross-correlation peak. Similar to the leading edge detection case, it can be shown that [6, 21]

$$\sigma^2_{\text{TDOA,xc}} \geq \frac{1}{8\pi B_{\text{RMS}} BT \xi_{eff}} \qquad (11.41)$$

where, ξ_{eff} is the effective SNR, computed as

$$\xi_{eff} = \frac{1}{\frac{1}{\xi_1} + \frac{1}{\xi_2} + \frac{1}{\xi_1 \xi_2}} \approx \min(\xi_1, \xi_2) \qquad (11.42)$$

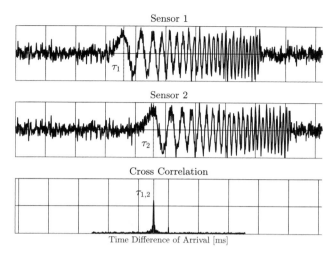

Figure 11.4 Illustration of cross-correlation TDOA processing. The received signals at sensors 1 and 2 are correlated against each other. The location of the peak indicates the TDOA between the two signals.

B is the passband bandwidth of the signal, T is the pulse duration, and B_{RMS} is the RMS bandwidth of the transmitted signal $S(f)$

$$B_{\text{RMS}} = \sqrt{\frac{\int_{-\infty}^{\infty} |S(f)|^2 f^2 df}{\int_{-\infty}^{\infty} |S(f)|^2 df}} \qquad (11.43)$$

Note that, for spectrally flat signals, this simplifies to $B_{\text{RMS}} = B/\sqrt{3}$. These terms increase the effective SNR when cross-correlation processing is used. Unlike the leading-edge case, the cross-correlation processing benefits from signals that use a large amount of pulse compression, such as LPI radar.

This equation is included in the provided MATLAB® code, under the function `tdoa.crossCorrError`.

Figure 11.5 plots the standard deviation σ_{TDOA} as a function of effective SNR for both peak detection and cross-correlation of a small bandwidth signal (a Link-16 style pulse with 5 MHz of bandwidth and 6-μs pulse duration,[4] which results in a time bandwidth product of $BT = 32$) and a large bandwidth signal

---
4 For details on Link-16 pulse structure, see [22].

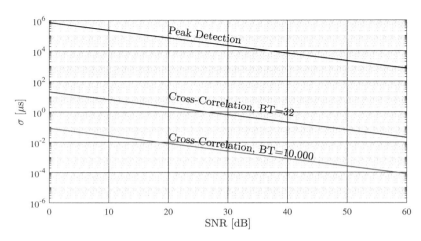

Figure 11.5 Plot of the TDOA error standard deviation as a function of SNR for (a) peak detection, (b) cross-correlation of a narrow bandwidth signal similar to a Link-16 pulse, and (c) cross-correlation of a wide bandwidth representative of a high resolution imaging pulse.

(a 1-GHz bandwidth pulse of a very high resolution imaging radar with 10-μs pulse length, which results in a time bandwidth product of $BT = 10,000$). In both cases, the signal is assumed to be spectrally flat, such that the RMS bandwidth is $B_{\text{RMS}} = B_s/\sqrt{3}$. Note that both of the cross-correlation variance curves are several orders of magnitude better than peak detection.

If two cross-correlation TDOA measurements share a common reference, as was the setup throughout this chapter, then their errors will also be correlated, as discussed in (11.6). If the two correlation measurements have the same error variance, then we can approximate the cross-correlation term as half of each error term's variance. However, exact calculation depends on the SNR of the reference sensor, relative to the two independent sensors.

11.4.3 Clock Synchronization

We've discussed errors in estimation of TDOA that arise from receiver noise, but not errors that arise from synchronization of the receivers. As systems become more accurate, leading to higher SNR levels and smaller variances in the estimate of single-sensor TOA, the clock drift between sensors begins to dominate. The total

TDOA error is a sum of both sources, since we can reasonably assume that receiver noise is uncorrelated with clock drift.

GPS has emerged as the *de facto* time synchronization standard, given its global availability and ease of implementation—there is no need to coordinate among sensors, just equip each with a GPS receiver and rely on its time signal. This leads to time synchronization errors on the order of 100–500 nanoseconds for nonkeyed receivers [23].[5] Recent tests by the U.S. Naval Observatory have demonstrated the ability to wirelessly synchronize a network of sensors spread over a 2.5-km area to within 200 picoseconds [24].

Although the extension of this to larger networks, particularly the 10–40-km baselines used in many fielded TDOA systems, will likely lead to increased errors, but this points to the *state of the possible* in time synchronization. We will show below, in Figure 11.6(b) that 100 ns timing accuracy supports a 10-km position error out to a range of 100 km for a reasonable geometry. Reduction to 1 ns (a five-fold increase over the demonstrated 200 ps lower limit) would scale the error result by three orders of magnitude, providing 100-m geolocation error (CEP_{50}) at 100-km standoff.

11.5 GEOLOCATION PERFORMANCE

A number of geolocation performance metrics were introduced in Chapter 9, including the CEP_{50}, and error ellipses. Straightforward computation requires the error covariance matrix, which is generally unavailable for TDOA algorithms, particularly for iterative methods such as the least squares and gradient descent approaches described here. Instead, most analysis of TDOA performance relies on Monte Carlo simulations and on statistical performance bounds. By far, the most popular is the CRLB.

Recall from Chapter 6 that the CRLB is a lower bound on the elements of the covariance matrix $\mathbf{C}_{\hat{\mathbf{x}}}$ and is given by the inverse of the FIM. In the general Gaussian case, with a measurement covariance matrix $\mathbf{C}_\rho$, the FIM is given as

$$\mathbf{F}(\mathbf{x}) = \mathbf{J}(\mathbf{x})\mathbf{C}_\rho^{-1}\mathbf{J}^T(\mathbf{x}) \tag{11.44}$$

where $\mathbf{J}(\mathbf{x})$ is the Jacobian matrix of the log likelihood function—in the case of TDOA, $\mathbf{J}(\mathbf{x})$ is defined in (11.10)—and, thus, the covariance matrix $\mathbf{C}_{\hat{\mathbf{x}}}$ of the

---

[5] Military receivers capable of processing the encrypted P(Y) code or upcoming M-code will likely have better performance.

estimated emitter position $\hat{\mathbf{x}}$ as a function of the true emitter position $\mathbf{x}$ is bounded as

$$\mathbf{C}_{\hat{\mathbf{x}}} \geq \left[\mathbf{J}(\mathbf{x}) \mathbf{C}_\rho^{-1} \mathbf{J}^T(\mathbf{x}) \right]^{-1} \qquad (11.45)$$

This bound must be solved numerically. An example is shown in Figure 11.6, under the assumption that each sensor's TDOA measurement has a standard deviation of 100 ns. The value plotted is the CEP_{50} computed from the CRLB. In this scenario, the three sensors are placed at even intervals along a circle that is 10 km in diameter, and the CRLB is computed for a 300 km x 300 km grid. The contours plotted show the CEP_{50} in units of kilometers. Note that, for this geometry, there are locations as far as 100 km from the center of the sensors, that have position errors below 10 km. This demonstrates the positioning accuracy possible via TDOA geolocation with 100-ns timing accuracy. Of course, there are other locations with errors in excess of 100 km, that are located only a few kilometers from one of the sensors, due to the poor geometry—for those locations, two of the sensors lie on a straight line, which traces an isochrone that is nearly parallel to the line of approach. The solution to this limitation is to add a fourth sensor.

The same scenario is repeated in Figure 11.6(11.6b) with the addition of a fourth sensor at the origin. This illustrates the impact of geometry, and the need to have $Ndim + 2$ sensors, rather than $Ndim + 1$, for stability.

Example 11.1 Four-Channel TDOA Solution

Consider an example four-channel TDOA system with three sensors equally spaced along a 10-km radius circle, centered on a reference sensor. The locations are plotted in Figure 11.7. In this example, there is an emitter roughly 50 km from the central reference center, marked with a triangle. We assume that each sensor achieves a standard deviation of 100 ns on TOA estimation. Generate solutions from a single set of timing measurements at the four sensors, using each of the approaches discussed in this chapter.

We begin by defining the sensor and source positions

```
% Sensor positions
r = 10e3*[1 1 1 0];
th = pi/2 + (0:3)*2*pi/3;
x0 = r.*[cos(th);sin(th)];

% Source position
th_s = 293.3;
xs = 50e3*[cos(th_s);sin(th_s)];
```

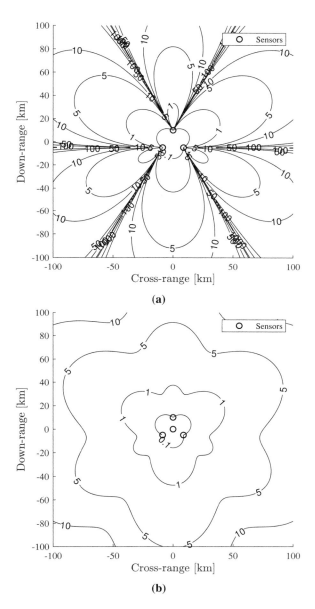

Figure 11.6 Calculation of RMSE for an emitter at any given point, based on the CRLB given the marked sensor positions and assumed standard deviation of 100 ns on all of the TDOA measurements: (a) three-sensor configuration, and (b) four-sensor configuration.

Next, we generate a set of range difference measurements

```
% Error
timingError = 1e-7;
rngStdDev = timingError * 3e8;
Ctdoa = timingError^2 * (1 + eye(3));
Crdoa = rngStdDev^2 * (1 + eye(3));

% Measurements
nMC = 1000;
dR = tdoa.measurement(x_sensor,x_source);
n = rngStdDev*(randn(3,nMC) + randn(1,nMC));
rho_MC = dR + n;   % Noisy range difference
```

Then, for each of the Monte Carlo iterations, we call the TDOA solutions, and compute the error, using the optional second output to collect each iteration of the solution, rather than just the final solution.

```
% Initial estimate in the right quadrant
x_init = 10e3*[cos(-pi/4);sin(-pi/4)];

for idx = 1:nMC
        rho = rho_MC(:,idx);
    [~,x_ls] = tdoa.lsSoln(x0,rho,diag(diag(Crdoa)),x_init,,numIters);
    [~,x_grad] = tdoa.gdSoln(x0,rho,Crdoa,x_init,alpha,beta,,numIters);
    x_chanHo = tdoa.chanHoSoln(x0,rho,Crdoa);

    % Compute Error
    thisErr_ls = xs-x_ls;            % [m]
    thisErr_grad = xs-x_grad;        % [m]
    thisErr_chanHo = xs-x_chanHo;    % [m]
end
```

The sensor and emitter positions are plotted in Figure 11.7, along with iterative position estimates using both the least squares and gradient descent algorithms presented, as well as the solution from the Chan-Ho algorithm presented above.

Note that for this example, the least squares solution converges to an estimate near the CRLB computed 50% error ellipse very quickly, within less than 10 iterations, while the gradient descent approach required more than 1,000 iterations. The Chan-Ho approach similarly lands close to the error ellipse. Figure 11.8 plots

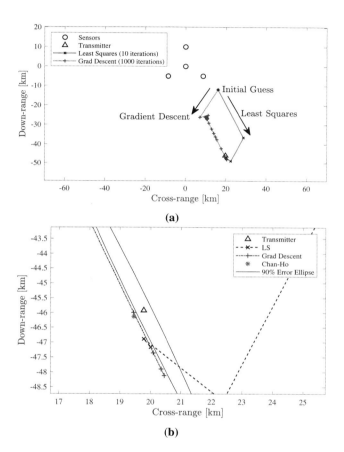

Figure 11.7 Graphical illustration of iterative solutions for Example 11.1: (a) Iterative solutions show the different paths taken by least square and gradient descent algorithms, and (b) close-up of emitter position shows the 90% error ellipse and final solutions for the three approaches modeled.

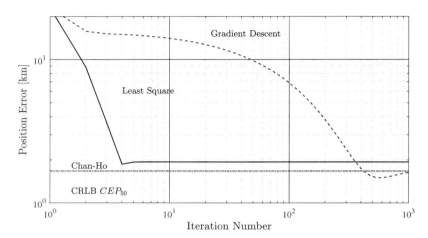

Figure 11.8 Plots of the CEP_{50} for position estimates in Example 11.1 (averaged over 10^4 Monte Carlo trials) for the three presented TDOA algorithms, as a function of iteration number, compared with the CEP_{50} computed from the CRLB. (Note that the Chan-Ho algorithm is not iterative, so it is plotted as a constant value.)

the CEP_{50} across iterations for each approach, averaged over 10^4 Monte Carlo trials, as well as for the CRLB in this scenario.[6]

11.6 LIMITATIONS

The analysis in this chapter has assumed perfect knowledge of the sensor locations. This is not practical in practice, especially for moving receivers, such as UAVs or ships at sea. While the algorithms in this chapter will still yield moderately accurate solutions, they will no longer achieve the expected performance bounds. Recent publications in the open literature have proposed algorithms for the consideration of sensor position errors (for stationary or moving sensors) [25].

Another problem in TDOA is the presence of multiple emitters. Multiple emitters cause a data fusion problem, with the potential that time of arrival estimates

---

[6] In computing the CEP_{50} for this example, we input the sum of the error covariance matrix $\mathbf{C}_{\hat{x}}$ as well as the bias term $\mathbf{bb}^T$. This is because the starting point was not varied, so the early stages of the iterative solutions are very similar. Calculating the CEP_{50} based on the covariance of the estimates alone would yield erroneously small error calculations.

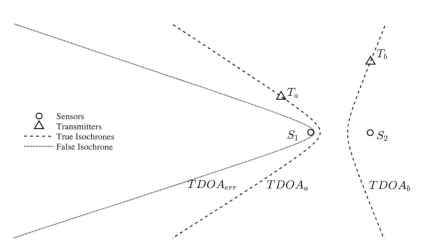

Figure 11.9 Drawing of the multisource problem, when pulses are misassociated, resulting in a false isochrone (dotted line).

will be incorrectly correlated (pulse A at sensor 1 may be compared with pulse B at sensor 2) as shown in Figure 11.9, or that hyperbolas are cross-correlated (TDOA solution from sensors 1 and 2 for pulse A compared with TDOA solution from sensors 1 and 3 for pulse B) as in Figure 11.10.

Typically, this problem is addressed by providing secondary information along with the time of arrival statistics, such as pulse descriptor words, which include various parameters of the received signal, such as pulse width, center frequency, bandwidth, modulation type, and possibly even AOA (if the sensor is equipped to perform that measurement). All of these secondary pieces of information assist the fusion engine in disambiguating pulse arrival times. Implementation of correlation TDOA, as opposed to leading edge of peak detection TOA, significantly improves the sorting problem, by providing the fusion engine with the full sampled signal, rather than a few estimated parameters. Nevertheless, there are algorithms proposed for the geolocation of multiple sources via TDOA, even without the addition of side channel information or cross-correlation receivers. Interested readers are referred to [26, 27].

Finally, but perhaps most importantly, TDOA relies on the target to emit signals, and it is vulnerable to EMCON. Simpler systems, such as those based on edge/peak detection of pulses at individual locations, also suffer from an inability to process continuous wave signals, and are less capable against LPI type signals

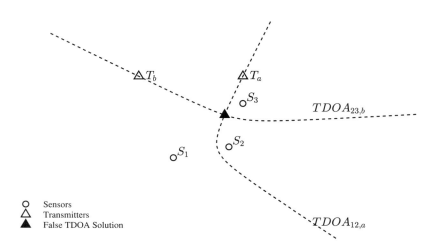

Figure 11.10 Drawing of the multisource problem, when isochrones are misassociated, resulting in a false location estimate.

with very large spreading bandwidth and low energy spectral density (W/Hz of spectrum). Cross-correlation receivers recover some of the performance loss against these emitters, but are still vulnerable to a target's radio silence. If the target is not emitting—or is not emitting in a band that is covered by the TDOA system—then it will not be detected or processed.

11.7 PROBLEM SET

11.1 Consider a three-channel TDOA system with sensors spaced 10 km apart along the x-axis, in a line. Each sensor generates time of arrival measurements with a standard deviation of 10 ns. Compute the CRLB for an emitter located at 75 km down-range, and 10 km cross-range, from this CRLB compute the RMSE.

11.2 Compute the standard deviation for peak detection TOA sensors, given a source located at 75 km range, with an ERP of 30 dBW and carrier frequency $f_0 = 9.5$ GHz. Assume the receiver has a modest 3-dBi array gain in the direction of the transmission, and has a receive bandwidth of $B_r = 100$ MHz, a noise figure $NF = 2$ dB, and a receive loss $L_r = 4$ dB. The transmitter altitude is $h_t = 10$ m and the receiver altitude is $h_r = 9$ km. If the transmit pulse has an

RMS bandwidth of $B_{\text{rms}} = 60$ MHz and pulse duration of $T = 100$ ms, then what will the standard deviation for a cross-correlation receiver be?

11.3 Repeat Problem 11.1 with the two-sensor performances computed in Problem 11.2.

11.4 Consider an example three-channel TDOA system with sensors spaced evenly along a circle of radius 25 km. Generate a single set of timing measurements with standard deviation of $\sigma_t = 150$ ns for each sensor from a source located 100 km down range. Apply the least square and gradient descent algorithms, with a maximum of 100 iterations for least square and 1,000 iterations gradient descent. Plot the intermediate and final solutions, to show how the algorithms converge. Overlay the true emitter position, 90% error ellipse from the CRLB, and source locations. Initialize the algorithms with a solution at 0 km cross-range, 2 km down-range (+y).

11.5 Attempt to compute the Chan-Ho solution for Problem 11.4; does it succeed? Why or why not?

11.6 Add a sensor at the origin to Problem 11.4 and repeat.

11.7 Add an offset sensor at $x = [50, 0]$ km to Problem 11.4 and repeat.

11.8 What is the CEP_{50} of the CRLB for Problems 11.4, 11.6, and 11.7.

References

[1] Jansky and Bailey, "The LORAN-C system of navigation," Atlantic Research Corporation, Tech. Rep., February 1962.

[2] W. R. Hahn, "Optimum signal processing for passive sonar range and bearing estimation," *The Journal of the Acoustical Society of America*, vol. 58, no. 1, pp. 201–207, 1975.

[3] J. Buisson and T. McCaskill, "Timation navigation satellite system constellation study," Naval Research Laboratory, Tech. Rep., 1972.

[4] N. Marchand, "Error distributions of best estimate of position from multiple time difference hyperbolic networks," *IEEE Transactions on Aerospace and Navigational Electronics*, no. 2, 1964.

[5] W. Crichlow, J. Herbstreit, E. Johnson, K. Norton, and C. Smith, "The Range Reliability and Accuracy of a Low Frequency Loran System," *Office of the Chief Signal Officer, The Pentagon, Washington, DC, Rept. ORS-P23, ASTIA Doc. AD*, vol. 52, p. 265, 1946.

[6] W. Hahn and S. Tretter, "Optimum processing for delay-vector estimation in passive signal arrays," *IEEE Transactions on Information Theory*, vol. 19, no. 5, pp. 608–614, 1973.

[7] W. J.E. Kaufmann, "Emitter Location with LES-8/9 Using Differential Time-of-Arrival and Differential Doppler Shift," MIT Lincoln Laboratory, Tech. Rep. TR-698 (Rev. 1), 2000.

[8] S. Boyd and L. Vandenberghe, *Convex optimization*. Cambridge, UK: Cambridge University Press, 2004.

[9] Y.-T. Chan and K. Ho, "A simple and efficient estimator for hyperbolic location," *IEEE Transactions on signal processing*, vol. 42, no. 8, pp. 1905–1915, 1994.

[10] Y. Zhou and L. Lamont, "Constrained linear least squares approach for TDOA localization: A global optimum solution," in *2008 IEEE International Conference on Acoustics, Speech and Signal Processing*, March 2008, pp. 2577–2580.

[11] L. Lin, H. So, F. K. Chan, Y. Chan, and K. Ho, "A new constrained weighted least squares algorithm for TDOA-based localization," *Signal Processing*, vol. 93, no. 11, pp. 2872 – 2878, 2013.

[12] F. Gustafsson and F. Gunnarsson, "Positioning using time-difference of arrival measurements," in *Acoustics, Speech, and Signal Processing, 2003. Proceedings.(ICASSP'03). 2003 IEEE International Conference on*, vol. 6. IEEE, 2003, pp. VI–553.

[13] K. C. Ho, "Bias reduction for an explicit solution of source localization using TDOA," *IEEE Transactions on Signal Processing*, vol. 60, no. 5, pp. 2101–2114, May 2012.

[14] B. Friedlander, "A passive localization algorithm and its accuracy analysis," *IEEE Journal of Oceanic engineering*, vol. 12, no. 1, pp. 234–245, 1987.

[15] J. Smith and J. Abel, "The spherical interpolation method of source localization," *IEEE Journal of Oceanic Engineering*, vol. 12, no. 1, pp. 246–252, 1987.

[16] J. Abel and J. Smith, "The spherical interpolation method for closed-form passive source localization using range difference measurements," in *ICASSP '87. IEEE International Conference on Acoustics, Speech, and Signal Processing*, vol. 12, Apr 1987, pp. 471–474.

[17] H. Boujemaa and M. Siala, "On a maximum likelihood delay acquisition algorithm," in *Communications, 2001. ICC 2001. IEEE International Conference on*, vol. 8. IEEE, 2001, pp. 2510–2514.

[18] V. Dizdarevic and K. Witrisal, "Statistical uwb range error model for the threshold leading edge detector," in *Information, Communications & Signal Processing, 2007 6th International Conference on*. IEEE, 2007, pp. 1–5.

[19] C. Steiner and A. Wittneben, "Robust time-of-arrival estimation with an energy detection receiver," in *Ultra-Wideband, 2009. ICUWB 2009. IEEE International Conference on*. IEEE, 2009.

[20] I. Sharp, K. Yu, and Y. J. Guo, "Peak and leading edge detection for time-of-arrival estimation in band-limited positioning systems," *IET Communications*, vol. 3, no. 10, pp. 1616–1627, October 2009.

References

[21] W. R. Hahn, "Optimum passive signal processing for array delay vector estimation," Naval Ordnance Laboratory, Tech. Rep., 1972.

[22] C.-H. Kao, "Performance analysis of a JTIDS/Link-16-type waveform transmitted over slow, flat Nakagami fading channels in the presence of narrowband interference," Ph.D. dissertation, Naval Postgraduate School, 2008.

[23] P. H. Dana and B. M. Penrod, "The role of GPS in precise time and frequency dissemination," *GPS World*, vol. 1, no. 4, pp. 38–43, 1990.

[24] E. Powers, A. Colina, and E. D. Powers, "Wide area wireless network synchronization using locata," *ION Publication*, pp. 9–20, 2015.

[25] K. C. Ho, X. Lu, and L. Kovavisaruch, "Source localization using TDOA and FDOA measurements in the presence of receiver location errors: Analysis and solution," *IEEE Transactions on Signal Processing*, vol. 55, no. 2, pp. 684–696, Feb 2007.

[26] L. Yang and K. C. Ho, "An approximately efficient TDOA localization algorithm in closed-form for locating multiple disjoint sources with erroneous sensor positions," *IEEE Transactions on Signal Processing*, vol. 57, no. 12, pp. 4598–4615, Dec 2009.

[27] M. Sun and K. C. Ho, "An asymptotically efficient estimator for TDOA and FDOA positioning of multiple disjoint sources in the presence of sensor location uncertainties," *IEEE Transactions on Signal Processing*, vol. 59, no. 7, pp. 3434–3440, July 2011.

Chapter 12

FDOA

FDOA is a high-precision emitter location technique, similar to TDOA, that is used to generate tracks of noncooperative emitting targets over time. Unlike TDOA, FDOA requires that either the emitter or (more commonly) the sensors are moving, and relies on differences in the relative velocity between the emitter and each target, which results in slightly different Doppler shift at each receiver.[1] This difference is measured and processed to estimate the emitter's position. It is more difficult to obtain accurate frequency measurements for FDOA, particularly given uncertainties in velocity and the dependence not just on the magnitude but on the direction of a sensor's motion. For these reasons, FDOA is less frequently used than TDOA. Nevertheless, it is a well understood and often studied technique.

In this chapter, we discuss the use of frequency measurements at a number of moving receiver stations to estimate an emitter's position. The approach closely matches that of TDOA discussed in Chapter 11, but has several important differences, which we will describe. Most modern formulations and solutions in the academic literature consider not FDOA in isolation, but in concert with TDOA, largely because this reduces the number of platforms required and the errors are often complementary. We will discuss the joint use of TDOA and FDOA in Chapter 13.

This chapter begins with a background on FDOA and formulation of the problem from a geographic perspective. We present several algorithms to solve for the unknown emitter position, given a set of FDOA measurements, and then discuss the expected errors in FDOA measurements and emitter geolocation.

---

[1] More formally, relative velocity causes dilation or compression of the time axis. For narrowband signals, this is approximated as a frequency shift.

12.1 BACKGROUND

Similar to TDOA, FDOA relies on a set of sensors, each estimating a parameter of the incoming signal, emitted by a source at some unknown position. The principal difference is that, with FDOA, the exact frequency of the received signal is measured at each receiver. If the receivers (or the source) are moving, then each of these receivers will measure a slightly different signal frequency, with the emitter's carrier frequency modulated by Doppler shift, representing the relative velocity of the emitter, as seen by each sensor. The changes in frequency between two receivers can be used to generate a partial solution, a curve along which the emitter lies. By comparing frequency at additional sensors, a position estimate can be formed. A graphic of this scenario is depicted in Figure 12.1.

In the TDOA case, the base time delay is an unknown that cannot be estimated directly, but through comparing the TDOA at a set of sensors, we can estimate the unknown bulk delay. Similarly, with FDOA, we cannot directly estimate the actual carrier frequency f_0, but through comparing the FDOA at each sensor, the Doppler shift at each sensor, and ultimately the emitter's position, can be estimated.

Due to its additional complexity, and issues generating frequency measurements with sufficient accuracy, FDOA is not as widely adopted or studied in the literature as TDOA. Nevertheless, formulations of the geometry of FDOA can be found in [1], and discussion of application to satellite constellations is found in [2]. Discussion of the underlying accuracy of estimating frequency difference (often referred to as *differential doppler*) and achievable emitter position accuracy trace to the 1970s, with most of the early work focused on sonar applications [3–8]. Since FDOA relies on moving sensors, with known velocity, it is most readily applied to satellite ELINT constellations.[2] This chapter does not discuss any of the intricacies of orbital mechanics, or specifics of accuracy for space-based sensors; interested readers are referred to [2, 11]. Instead, we focus on the generic problem using 2-D coordinates and frequency measurements.

Also, similar to TDOA and triangulation, estimation of emitter position is greatly improved by applying tracking approaches. For simplicity, we consider single snapshot position estimation. Most of the available literature on the subject utilizes multiple modalities (such as TDOA/FDOA, or TDOA/FDOA/AOA). Interested readers are referred to [10, 12, 13].

---

2   The problem can be formulated as easily for stationary receivers and a moving emitter, but with the added complication of estimating the emitter's velocity. Typically, in this case, TDOA and FDOA are jointly leveraged to compensate for the extra unknowns [9, 10].

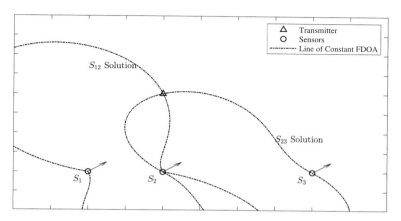

Figure 12.1 Example FDOA scenario showing three moving sensors and the FDOA solution for two pairs.

12.2 FORMULATION

Our formulation is based loosely on the signal models in [14–16] We begin with a transmitted signal $s(t)$ that experiences attenuation, delay, and doppler shift.

$$y_i(t) = \alpha_i s(t - \tau_i) e^{j f_i (t - \tau_i)} + n_i(t) \qquad (12.1)$$

where τ_i is the time delay to the ith receiver, and f_i is the Doppler shift induced by relative motion between the emitter and the receiver. $n_i(t)$ is the noise term. We define the range to the ith sensor, and the range rate (change in range over time) for an emitter at position $\mathbf{x}$ with velocity $\mathbf{v}$

$$R_i(\mathbf{x}) \triangleq |\mathbf{x} - \mathbf{x}_i| \qquad (12.2)$$

$$\dot{R}_i(\mathbf{x}, \mathbf{v}) \triangleq \frac{(\mathbf{v}_i - \mathbf{v})^T (\mathbf{x} - \mathbf{x}_i)}{R_i(\mathbf{x})} \qquad (12.3)$$

From this geometry, the time delay and Doppler shift are given as

$$\tau_i = \frac{R_i}{c} \qquad (12.4)$$

$$f_i = \frac{f_0}{c} \dot{R}_i \qquad (12.5)$$

12.2.0.1 IsoDoppler Contours

Just as, in TDOA, each pair of sensor's results in a set of isochrones, each pair of sensors in an FDOA has a set of isodoppler contours, or lines of constant doppler difference. This is illustrated for a test case in Figures 12.2. The frequency difference between two sensors is defined

$$f_{m,n}(\mathbf{x}, \mathbf{v}) = f_m(\mathbf{x}, \mathbf{v}) - f_n(\mathbf{x}, \mathbf{v}) \tag{12.6}$$

$$= \frac{f_0}{c} \left[\frac{(\mathbf{v}_m - \mathbf{v})^T (\mathbf{x} - \mathbf{x}_m)}{\|\mathbf{x} - \mathbf{x}_m\|_2} - \frac{(\mathbf{v}_n - \mathbf{v})^T (\mathbf{x} - \mathbf{x}_n)}{\|\mathbf{x} - \mathbf{x}_n\|_2} \right] \tag{12.7}$$

For simplicity, we assume that $\mathbf{v} = \mathbf{0}$. If that is not the case, then $\mathbf{v}$ represents an additional unknown (actually, one per spatial dimension). To estimate $\mathbf{v}$ in addition to $\mathbf{x}$, additional information must be collected via either additional sensors, or additional modalities (e.g., TDOA or AOA) as discussed in Chapter 13.

We plot isodoppler contours for two simple scenarios in Figure 12.2. In the first case, the velocities are aligned with the axis of the two sensors and the contours resemble those of a magnetic field. Directly between the sensors is the peak frequency difference (one is positive, while the other is negative), and the contours gradually progress toward the point of equality in the regions to the left and right of the two sensors. In the second case, the velocities are perpendicular to the central axis. The contours are more complex, but relatively simple to understand. In each quadrant, there is one sensor that is off-axis and has a smaller doppler shift (S_0 for the first and fourth, S_1 for the second and third). In the upper quadrants, the one more closely aligned has a larger positive shift, while in the lower quadrants it has a larger negative shift. In other, more realistic, scenarios, the contours will be much more complex and difficult to intuitively explain, but they follow these same principles.

12.3 SOLUTION

Consider the frequency difference measurements from sensors 1 through $N - 1$, as compared to a common reference (sensor N). We compute the range rate difference from these Doppler shifts between the mth and nth sensors $\dot{R}_{m,n}(\mathbf{x})$

$$\dot{R}_{m,n}(\mathbf{x}) = \frac{c}{f_0} f_{m,n}(\mathbf{x}) = \left[\frac{\mathbf{v}_m^T (\mathbf{x} - \mathbf{x}_m)}{\|\mathbf{x} - \mathbf{x}_m\|_2} - \frac{\mathbf{v}_n^T (\mathbf{x} - \mathbf{x}_n)}{\|\mathbf{x} - \mathbf{x}_n\|_2} \right] \tag{12.8}$$

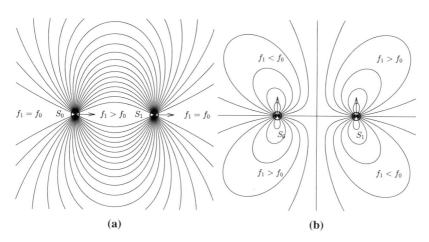

Figure 12.2 Iso-Doppler contours trace lines of constant frequency *difference* between two receivers for a stationary emitter at various locations: (a) Both sensors moving in the $+x$ direction and (b) both sensors moving in the $+y$ direction.

and collect the full vector

$$\dot{\mathbf{r}}(\mathbf{x}) = \left[\dot{R}_{1,N}(\mathbf{x}), \ldots, \dot{R}_{N-1,N}(\mathbf{x}) \right]^T \tag{12.9}$$

Noisy measurements of the range-rate difference vector are given

$$\dot{\boldsymbol{\rho}} = \dot{\mathbf{r}}(\mathbf{x}) + \mathbf{n} \tag{12.10}$$

where $\mathbf{n}$ is a Gaussian random vector with covariance $\mathbf{C}_{\dot{\mathbf{r}}}$. We assume all frequency measurements are independent, with variance σ_m^2 (Hz^2). The covariance matrix can thus be written by converting from Doppler frequency difference to range-rate difference.

$$\mathbf{C}_{\dot{\mathbf{r}}} = \frac{c^2}{f_0^2} \begin{bmatrix} \sigma_1^2 + \sigma_N^2 & \sigma_N^2 & \cdots & \sigma_N^2 \\ \sigma_N^2 & \sigma_2^2 + \sigma_N^2 & \cdots & \\ \vdots & & \ddots & \vdots \\ \sigma_1^2 & \cdots & & \sigma_{N-1}^2 + \sigma_N^2 \end{bmatrix} \tag{12.11}$$

$\mathbf{C}_{\dot{\mathbf{r}}}$ is given in units of m²/s². This is the same formulation as we saw for the measurements $\dot{\boldsymbol{\rho}}$ of the range difference vector $\dot{\mathbf{r}}(\mathbf{x})$. Thus, the PDF and log-likelihood function are similarly given as

$$f_\mathbf{x}(\dot{\boldsymbol{\rho}}) = (2\pi)^{-\frac{(N-1)}{2}} |\mathbf{C}_{\dot{\mathbf{r}}}|^{-1/2} e^{-\frac{1}{2}(\dot{\boldsymbol{\rho}} - \dot{\mathbf{r}}(\mathbf{x}))^T \mathbf{C}_{\dot{\mathbf{r}}}^{-1}(\dot{\boldsymbol{\rho}} - \dot{\mathbf{r}}(\mathbf{x}))} \qquad (12.12)$$

and

$$\ell(\mathbf{x}|\dot{\boldsymbol{\rho}}) = -\frac{1}{2}(\dot{\boldsymbol{\rho}} - \dot{\mathbf{r}}(\mathbf{x}))^T \mathbf{C}_{\dot{\mathbf{r}}}^{-1}(\dot{\boldsymbol{\rho}} - \dot{\mathbf{r}}(\mathbf{x})) \qquad (12.13)$$

The log-likelihood can be evaluated with the function `fdoa.loglikelihood` in the provided MATLAB® package, and a noise-free measurement can be quickly generated with the function `fdoa.measurement`.

12.3.1 Maximum Likelihood Estimate

The maximum likelihood estimate is found by maximizing the log-likelihood of $\dot{\boldsymbol{\rho}}$ given the unknown emitter position $\mathbf{x}$. We take the derivative of the log likelihood with respect to the unknown sensor position $\mathbf{x}$, given as

$$\nabla_\mathbf{x} \ell(\mathbf{x}|\dot{\boldsymbol{\rho}}) = \mathbf{J}(\mathbf{x}) \mathbf{C}_{\dot{\mathbf{r}}}^{-1} (\dot{\boldsymbol{\rho}} - \dot{\mathbf{r}}(\mathbf{x})) \qquad (12.14)$$

where $\dot{\mathbf{J}}(\mathbf{x})$ is the Jacobian matrix of $\dot{\mathbf{r}}(\mathbf{x})$

$$\mathbf{J}(\mathbf{x}) = \left[\nabla_\mathbf{x} \dot{R}_{1,N}, \ldots, \nabla_\mathbf{x} \dot{R}_{N-1,N} \right] \qquad (12.15)$$

and $\nabla_\mathbf{x} \dot{R}_{n,N}^T$ is the gradient of the range-rate difference to the emitter between the nth sensor and the reference sensor

$$\nabla_\mathbf{x} \dot{R}_{n,N} = (\mathbf{I} - \mathbf{P}_n(\mathbf{x})) \frac{\mathbf{v}_n}{\|\mathbf{x}_n - \mathbf{x}\|_2} - (\mathbf{I} - \mathbf{P}_N(\mathbf{x})) \frac{\mathbf{v}_N}{\|\mathbf{x}_N - \mathbf{x}\|_2} \qquad (12.16)$$

where

$$P_n(\mathbf{x}) = \frac{(\mathbf{x} - \mathbf{x}_n)(\mathbf{x} - \mathbf{x}_n)^T}{\|\mathbf{x} - \mathbf{x}_n\|_2^2} \qquad (12.17)$$

is the projection matrix onto the line of sight between the nth sensor and the emitter. Intuitively, this states that the range rate difference between sensors n and N is sensitive primarily to the amount of each sensor's velocity that is orthogonal to that sensor's line of sight, and the distance between that sensor and the emitter. It also indicates that performance is maximized (i.e., the gradient is most sensitive to changes in x) when $\mathbf{v}_n$ is orthogonal to $\mathbf{x} - \mathbf{x}_n$ and/or $\mathbf{v}_N$ is orthogonal to $\mathbf{x} - \mathbf{x}_N$, since those orientations will lead to the largest change in $\dot{R}_{n,N}$ for a given change in x.

The Jacobian can be directly computed with `fdoa.jacobian` in the provided MATLAB® package, and the ML estimate can be approximated with a brute force search via `fdoa.mlSoln`.

We plot an example of the brute force ML estimate in Figure 12.3 for a noise-free test case. For illustration, we first show solutions from individual sensor pairs, and then the full solution. Sensors are marked in all three plots with an 'x,' and the source is marked with an 'o.' The ML estimate is given at the intersection of the contours.

12.3.2 Iterative Least Squares Solution

The most straightforward approach to solving (12.13) is to posit an estimated position, and then use a Taylor series approximation to update it. This is the same approach described for iterative least squares in Chapters 10 and 11.

The gradient of each range rate difference is collected in the Jacobian matrix $\mathbf{J}(\mathbf{x})$ (12.15). Using this, we can express the linear approximation to (12.9) about some position estimate $\mathbf{x}^{(i)}$, and with offset $\Delta \mathbf{x}^{(i)}$:

$$\dot{\mathbf{r}}\left(\mathbf{x}^{(i)} + \Delta\mathbf{x}^{(i)}\right) = \dot{\mathbf{r}}\left(\mathbf{x}^{(i)}\right) + \mathbf{J}^T\left(\mathbf{x}^{(i)}\right)\Delta\mathbf{x}^{(i)} \qquad (12.18)$$

The rest of the derivation follows identically from that in Section 11.3.2, with several variables swapped for their FDOA equivalents. Notably, the range difference vector $\mathbf{r}$ is replaced with range-rate difference $\dot{\mathbf{r}}$ and corresponding measurement vectors, covariance matrices, and Jacobian matrices.

Recall that, in order to guarantee that this solution is not underdefined, the Jacobian matrix $\mathbf{J}$ must not have more columns than rows. In other words, the number of unique FDOA pairs $(N-1)$ must be greater than or equal to the number of spatial dimensions. If the emitter has unknown velocity as well, then the number of unique FDOA pairs $(N-1)$ must be greater than twice the number of spatial dimensions. This is a necessary condition, but not a sufficient one. It is possible for $\mathbf{J}$ to be underdefined even if it contains more columns than rows; this is the case when one column is a linear combination of the others.

Just as in the TDOA case, the least square solution is not guaranteed to be optimal, only to be closer than the initial estimate $\left(\mathbf{x}^{(i)}\right)$ under the assumption

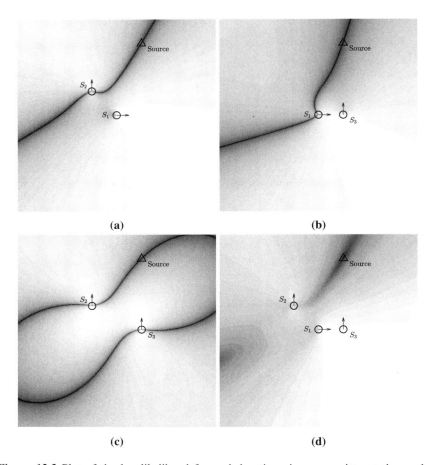

Figure 12.3 Plot of the log likelihood for each location given an emitter at the marked location (marked by an △) as seen by the FDOA sensors (marked by an o). All plots are on a log scale, with 80 dB of contrast: (a) sensors 1 and 2, (b) sensors 1 and 3, (c) sensors 2 and 3, and (d) all three sensors.

that $\mathbf{x}^{(i)}$ is sufficiently close to the true position that the problem is indeed locally convex. In order to obtain an accurate solution, the process must be repeated in an iterative fashion. An implementation of this algorithm is included in the provided MATLAB® code, under the function `fdoa.lsSoln`.

Recall from Chapter 10 that the principal limitations of this approach are the need for repeated calculations to reach convergence and the need for a sufficiently accurate initial position estimate. It is very important to understand that "sufficiently accurate" is ill defined, and there is thus little guarantee that an initial position estimate used to seed this approach has satisfied the requirement for local convexity.

Furthermore, this technique suffers from robustness concerns, namely that it is possible for the estimate to diverge wildly from the true emitter location. To improve numerical stability, we recommend supplying the algorithm with only the diagonal elements of $\mathbf{C}_{\dot{\rho}}$, rather than the full covariance matrix

```
C_ls = diag(diag(C));
```

Constraints and other optimizations are often necessary to improve accuracy. There is a rich field of research developing new least squares based solutions to the FDOA problem. See, for example [17, 18].

12.3.3 Gradient Descent Algorithms

We can apply the gradient descent algorithm from Section 11.3.3 to FDOA geolocation directly, by substituting the FDOA definitions for the Jacobian $\mathbf{J}(\mathbf{x})$ and offset vector $\mathbf{y}(\mathbf{x})$. We omit the formulation here for brevity, but note that an implementation can be found in the provided MATLAB® package, with the function `fdoa.gdSoln`.

Perhaps the most important feature of gradient descent algorithms is that the update equation does not contain a matrix inversion, so it is much faster to compute than the iterative least square solution.

The principal downside is still the need for a suitably accurate initial estimate, such that convexity of the objective function can be assumed.

12.3.4 Other Approaches

Recently, Cameron derived a special solution algorithm for two FDOA receivers that leverages samples over time (assuming a stationary receiver), and randomly samples triplets of measurements, forms a position fix from each triplet, and then searches for consensus among the solutions [19, 20]. This is an application of the

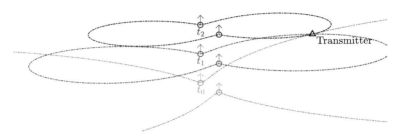

Figure 12.4 Illustration of two-sensor FDOA from moving platforms, with multiple measurements taken at different points in time.

RANSAC (random sampling and consensus) algorithm [21] to FDOA, and relies on a polynomial representation of the FDOA system of equations, similar to that outlined in [9, 14, 22].

The broader approach of using repeated measurements from two sensors leverages the stationary emitter assumption to generate n FDOA measurements at n different points in time, as shown in Figure 12.4. This approach is very useful for FDOA, since the emitter is already frequently assumed to be stationary, and it minimizes the number of sensor platforms required (from four for 3-D FDOA to two). The only drawback is the need for the emitter to be active for a longer observation interval, to allow the moving sensors to traverse a larger baseline (ideally >10 km for long-range geolocation), and the added challenge of fusing measurements if the environment is densely populated.

Mušicki and Koch [16] point out that while the FDOA measurements are Gaussian, when they are projected onto spatial coordinates, they are decidedly non-Gaussian (since $\dot{r}(x)$ is non-linear). As such, using Gaussian assumptions to derive the ML estimate and CRLB, while convenient, is not strictly accurate. They argue that proper estimation within the observation space requires nonlinear information fusion of the various FDOA measurements, which they perform with a Gaussian measurement mixture (GMM) algorithm. Interested readers are referred to [16] for details. Their arguments notwithstanding, we will proceed with CRLB analysis in this text, given its ease of derivation and calculation. However, we acknowledge its shortcomings.

12.4 FDOA ESTIMATION

A critical component of FDOA is the estimation of the frequency difference between each pair of sensors. Similar to TDOA, this can be achieved by first estimating frequency at each sensor, and then comparing those estimates, or by direct comparison of the received signals. The latter is much more accurate, but comes with the additional burden of transmitting raw sensor data.[3] In this section, we briefly review each approach, and provide references.

12.4.1 Frequency of Arrival Estimation

Estimation of signal frequency is a well understood problem in EW literature, and has a broad array of solutions. Adamy lists several in Chapter 6 of [24], and we briefly discuss the frequency accuracy of some receiver types in Chapter 5. Interested readers are also directed to [25] for a review of the history and development of instantaneous frequency measurement techniques, and [26, 27] for a comprehensive discussion for general signals. In general, frequency resolution is limited by the Nyquist sampling limit

$$\sigma_f \geq 1/T_s \tag{12.19}$$

where T_s is the observation interval. While the Nyquist theorem limits the *resolvability* of closely spaced signals, it is not the limit of estimation accuracy, which is more properly given by the CRLB on estimation error. The CRLB for frequency estimation of an unknown sinusoid is given as [28]

$$\sigma_f^2 \geq \frac{3}{\pi^2 t_s^2 M(M^2 - 1)\xi} \tag{12.20}$$

where t_s is the sampling period (sec),[4] M is the number of samples, and ξ is the per-sample SNR. A similar result is also derived in [29] in the presence of amplitude and phase nuisance parameters.

---

3 For a discussion of how to compress those signals, to minimize communications needs, see [23].
4 Readers will note that the formula in [28] does not contain the term t_s; this is because the frequency reported therein is in digital frequency units (1/sample), whereas we seek the frequency accuracy in Hertz (1/sec).

12.4.2 FDOA Estimation

The classical result for Cramer-Rao bound on FDOA between two receivers is given by Stein in 1981 [4]. If we assume that time delay is negligible, then the frequency difference estimate is given as [4, 10, 15, 16]

$$\sigma_{fd}^2 \geq \frac{1}{4\pi^2 T_{rms}^2 B_r T_s \xi_{eff}} \tag{12.21}$$

where $B_r T_s$ is the time-bandwidth product of the transmitted signal (limited by the receiver bandwidth and observation interval, respectively), T_{rms} is the RMS signal duration (analogous to the RMS bandwidth used in Chapter 11)

$$T_{rms}^2 = \frac{\int_{-\infty}^{\infty} t^2 |s(t)|^2 \, dt}{\int_{-\infty}^{\infty} |s(t)| \, dt} \tag{12.22}$$

and ξ_{eff} was defined in Chapter 11 as the effective SNR from combining the two measurements

$$\xi_{eff} = 2 \left[\frac{1}{\xi_1} + \frac{1}{\xi_2} + \frac{1}{\xi_1 \xi_2} \right]^{-1} \tag{12.23}$$

For a signal with a constant envelope of duration T_s, the RMS duration simplifies to

$$T_{rms}^2 = \frac{\int_0^{T_s} t^2 \, dt}{\int_0^{T_s} 1 \, dt} = \frac{T_s^2}{3} \tag{12.24}$$

and thus the bound can be rewritten

$$\sigma_{fd}^2 \geq \frac{3}{4\pi^2 B_r T_s^3 \xi_{eff}} \tag{12.25}$$

To compare the two methods, we plot them both as a function of SNR (ξ_{eff}) for two signals, an unmodulated pulse (with $B_r T_s = 1$) and a modulated pulse (with $B_r T_s = 1,000$). In both cases, the frequency difference estimation accuracy is approximately 6 dB better than raw frequency analysis. If the bandwidth is available to support data exchange, it is obviously better to perform frequency difference estimation.

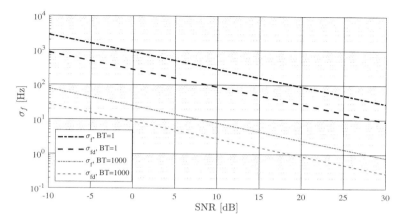

Figure 12.5 Comparison of frequency estimation (σ_f) and frequency difference estimation (σ_{fd}) accuracy for two separate signals; an unmodulated pulse (BT=1) and a modulated pulse (BT=1,000). In each case, the observation interval and pulse length are 1 ms, the bandwidth is varied.

12.4.3 Limitations of Frequency Estimation

Note that, for some waveforms, there is an inherent ambiguity between delay and Doppler. Thus, if delay is not assumed to be negligible (or otherwise known), it is impossible to reliably estimate frequency difference. This is most notably the case for linear frequency modulated (LFM) chirps, which sweep across frequency during a pulse, and are widely used by radar systems [30]. If the target signal is a chirp, it will be difficult to accurately measure delay and doppler. Similarly, for more general wideband systems, the Doppler shift is not a simple frequency shift, but rather a dilation of the time axis, and more complicated algorithms are required to accurately estimate it. Interested readers are referred to [31, 32].

12.5 GEOLOCATION PERFORMANCE

Similar to the case of TDOA, we use the CRLB to bound geolocation performance as a function of geometry and individual sensor performance. The CRLB from (11.45) can be directly applied, with a simple replacement of comparable variables. The covariance matrix $\mathbf{C}_\rho$ is replaced with $\mathbf{C}_{\dot{\rho}}$, and the jacobian $\mathbf{J}$

from (11.10) is replaced with its FDOA counterpart (12.15).

$$\mathbf{C}_{\hat{\mathbf{x}}} \geq \left[\mathbf{J}(\mathbf{x}) \mathbf{C}_{\hat{\rho}}^{-1} \mathbf{J}^{T}(\mathbf{x}) \right]^{-1} \quad (12.26)$$

This bound must be solved numerically. An example is shown in Figure 12.6(a), under the assumption that each sensor's FDOA measurement has a standard deviation of 10 Hz. The value plotted is the CEP_{50} computed from the CRLB. In this scenario, the three sensors are placed at even intervals along a circle that is 10 km in diameter, with 100 m/s velocity, and the CRLB is computed for a 300 km x 300 km grid. The contours plotted show the CEP_{50} in units of kilometers. Note that, for this geometry, there are locations as far as 100 km from the center of the sensors, that have a position errors below 50 km. Note that the lobing structure of the contours resemble that of TDOA in Figure 11.6(a), but the accuracy overall is much worse. This shows that even an aggressive 10-Hz frequency estimation accuracy is not as powerful as a 100-ns temporal estimation accuracy with TDOA, for the geometry and sensor velocity shown. If the emitter is at a higher frequency (say, 10 GHz or above) this will improve, but the sensors have no control over what frequency an emitter will utilize. Additionally, performance can be improved by increasing the sensor velocities, such that the induced Doppler shifts are increased, but that is limited by the host vehicle carrying the sensor. In general, TDOA techniques are more accurate.

The same scenario is repeated with a change to the sensor velocities in Figure 12.6(b), which are now aligned in the +x direction, as opposed to traveling away from each other. In this case, the performance is not vastly improved, but the lobing structure is completely different. Performance is guaranteed over a much broader range (to the side of the formation, in the +y and -y directions) while targets in front of or behind (in the +x and -x directions) are much more difficult to accurately localize. This is caused by the shape of the isodoppler contours, which are fairly flat in front of and behind the formation (when all sensors have the same direction of travel), and the impact of fully aligned sensor velocities.

The same scenario is repeated in Figures 12.6(c, d) with the addition of a fourth sensor at the origin. This illustrates the impact of geometry, the need to have $Ndim + 2$ sensors, rather than $Ndim + 1$, for stability.

Example 12.1 Four-Channel FDOA Solution

Consider an example four-channel FDOA system with three sensors equally spaced along a 10-km radius circle, centered on a reference sensor. Each is moving away

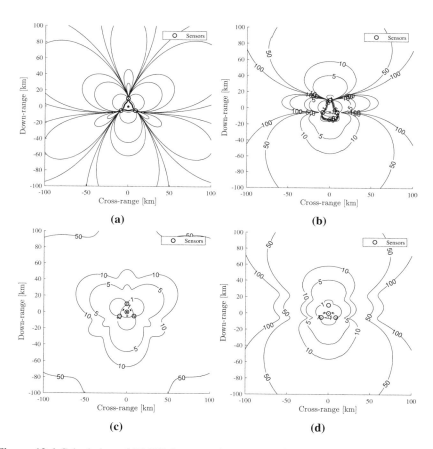

Figure 12.6 Calculation of RMSE for an emitter at any given point, based on the CRLB given the marked sensor positions and assumed standard deviation of 10 Hz on all of the FDOA measurements, carrier frequency of 1 GHz, and sensor velocity of 100 m/s: (a) three-sensor configuration, with sensors moving radially out (b) three-sensor configuration, with all sensors moving in $+x$ direction, (c) four-sensor configuration with sensors moving radially out, and (d) four-sensor configuration with sensors moving in $+x$ direction.

from the origin at 100 m/s (with the center reference center moving in the +x direction). The locations are plotted in Figure 12.7. In this example, there is an emitter roughly 50 km from the central reference center, marked with a triangle. We assume that each sensor achieves a standard deviation of 3 Hz on frequency estimation. Generate solutions from a single set of timing measurements at the four sensors, using each of the approaches discussed in this chapter.

We begin by defining the sensor and source positions

```
% Sensor positions
r = 10e3*[1 1 1 0];
th = pi/2 + (0:3)*2*pi/3;
x0 = r.*[cos(th);sin(th)];
v0 = 100.*[cos(th);sin(th)]; v0(:,end) = 100*[1;0];

% Source position
th_s = 4.9363;
xs = 50e3*[cos(th_s);sin(th_s)];
```

Next, we generate a set of range rate difference measurements

```
% Error
freqError = 3;
f0 = 1e9;
rngRtStdDev = freqError* 3e8 / f0;
Cfdoa = freqError^2*(1 + eye(3));
Crrdoa = rngRateStdDev^2*(1 + eye(3));

% Measurements
nMC = 10000;
r = fdoa.measurement(x0,v0,xs);
n = rngStdDev*(randn(3,nMC) + randn(3,nMC)); % Noisy range difference
rho_dot_mc = dR + n;
```

Then, for each of the Monte Carlo iterations, we call the FDOA solutions, and compute the error, using the optional second output to collect each iteration of the solution, rather than just the final solution.

```
x_init = 10e3*[cos(th_s); sin(th_s)]; % Initial estimate
for idx = 1:nMC
    rho_dot = rho_dot_mc(:,idx);
    [~,x_ls] = fdoa.lsSoln(x0,rho_dot,diag(diag(Crrdoa)),x_init,[],
        numIters);
```

```
    [~,x_grad] = fdoa.gdSoln(x0,v0,rho_dot,Crrdoa,x_init,alpha,beta,[],
        numIters);

    % Compute Error
    thisErr_ls = xs-x_ls;      % [m]
    thisErr_grad = xs-x_grad;  % [m]
end
```

The sensor and emitter positions are plotted in Figure 12.7, along with iterative position estimates using both the least squares and gradient descent algorithms presented.

Analysis of Figure 12.7 shows that, based on the source position and initial estimate used here, the gradient descent algorithm oscillates around an isodoppler contour that leads to the target, until it falls into that valley and travels smoothly along the contour.

Figure 12.8 plots the CEP_{50} across iterations for each approach, averaged over 10^4 Monte Carlo trials, as well as for the CRLB in this scenario.[5]

Note that for this example, the least squares solution converges to an estimate near the CRLB computed 50% error ellipse very quickly, within less than 10 iterations, while the gradient descent approach required more than 1,000 iterations. Also, gradient descent appears to settle on a steady state error close to the CRLB, while least square appears to outperform it. One possible explanation is that, due to the nonlinearity of $\dot{r}(x)$, the error term in $\hat{x}$ is non-Gaussian, and thus the CRLB is no longer a strict bound. It is also likely that this outperformance is sensitive to the initial position estimate and the sensor arrangement.

12.6 LIMITATIONS

FDOA relies on fusion of information from multiple sources, as does TDOA and AOA. One of the principal limitations of multisensor information fusion is the reliability of that information. If an adversary can contaminate information, or inject false reports, into the data stream, it can cripple geolocation performance. Some aspects of this problem, and how to respond to it, are introduced in recent literature [33], but remains an open (and vigorously studied) field.

---

[5] In computing the CEP_{50} for this example, we input the sum of the error covariance matrix $C_{\hat{x}}$ as well as the bias term bb^T. This is because the starting point was not varied, so the early stages of the iterative solutions are very similar. Calculating the CEP_{50} based on the covariance of the estimates alone would yield erroneously small error calculations.

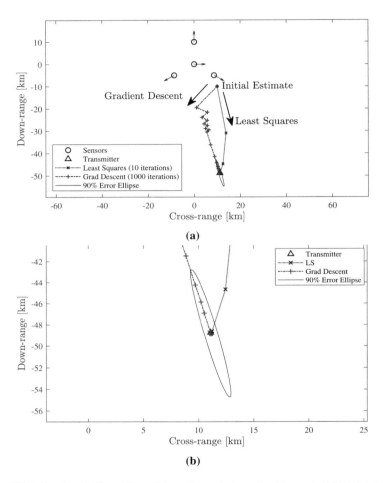

Figure 12.7 Graphical illustration of iterative solutions for Example 12.1: (a) iterative solutions show the different paths taken by least square and gradient descent algorithms, and (b) close-up of emitter position shows the 90% error ellipse, and final solutions for the two approaches modeled.

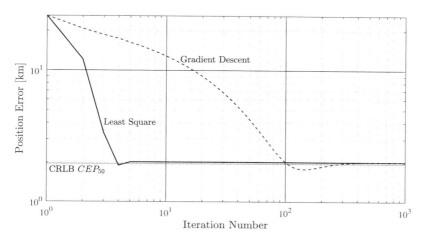

Figure 12.8 CEP_{50} for position estimates in Example 12.1 (averaged over 10^4 Monte Carlo trials) for the two presented FDOA algorithms, as a function of iteration number, compared with the CEP_{50} computed from the CRLB.

All of the other limitations of TDOA apply equally to FDOA as well, including those of multiple targets, data association, false position estimates, and, most importantly, vulnerability to EMCON.

12.7 PROBLEM SET

12.1 Generate a noisy set of measurements based on Example 12.1, but with all four sensors traveling in the $+y$ direction, the source at 50 km from the origin, at $45°$ CW from $+y$ direction. Compute position estimates using both the iterative least square and gradient descent algorithms; allow them to run for as many as 100 and 300 iterations, respectively. Initialize both algorithms with an estimate 20 km from the origin, at $20°$ CW from the $+y$ axis. Comment on the performance.

12.2 Compute the CRLB and the RMSE. Repeat Problem 12.1 for 100 random Monte Carlo trials, and compute the RMSE achieved by each solution. Compare to the CRLB.

12.3 Repeat Problem 12.1, but reduce sensor velocity to 50 m/s. Comment on the change in performance.

12.4 Repeat Problem 12.2, but reduce sensor velocity to 50 m/s. Comment on the change in performance.

12.5 Consider a two-ship spaced $x = 10$ km apart, moving in the $+y$ direction (side-to-side) with $v = 500$ m/s. A set of FDOA measurements are taken every 10 seconds for a total of $N = 7$ measurements. Plot the CRLB contour for emitters as far as 250 km from the center of the two sensors at the first time step, with a center frequency $f_0 = 3$ GHz; assume sensor accuracy is $\delta_f = 5$ Hz. Note that fdoa.computeCRLB has an optional argument refIdx that can be used to define individual FDOA sensor pairs. Type help fdoa.computeCRLB for details on how to use it.

12.6 Repeat Problem 12.5, but change the velocity vector on the two sensors, such that the angle between their headings is $10°$ (one at $+5°$ from the y-axis and one at $-5°$). Comment on the changes.

References

[1] D. Torrieri, "Statistical theory of passive location systems," Army Materiel Development and Readiness Command, Countermeasures/Counter-Countermeasures Office, Tech. Rep. ADA128346, 1983.

[2] W. J.E. Kaufmann, "Emitter Location with LES-8/9 Using Differential Time-of-Arrival and Differential Doppler Shift," MIT Lincoln Laboratory, Tech. Rep. TR-698 (Rev. 1), 2000.

[3] P. M. Schultheiss and E. Weinstein, "Estimation of differential doppler shifts," *The Journal of the Acoustical Society of America*, vol. 66, no. 5, pp. 1412–1419, 1979.

[4] S. Stein, "Algorithms for ambiguity function processing," *IEEE Transactions on Acoustics, Speech, and Signal Processing*, vol. 29, no. 3, pp. 588–599, June 1981.

[5] P. C. Chestnut, "Emitter location accuracy using TDOA and differential Doppler," *IEEE Transactions on Aerospace and Electronic Systems*, no. 2, pp. 214–218, 1982.

[6] S. Stein, "Differential delay/doppler ml estimation with unknown signals," *IEEE Transactions on Signal Processing*, vol. 41, no. 8, pp. 2717–2719, 1993.

[7] B. Friedlander, "On the cramer-rao bound for time delay and doppler estimation (corresp.)," *IEEE Transactions on Information Theory*, vol. 30, no. 3, pp. 575–580, 1984.

[8] M. Wax, "The joint estimation of differential delay, doppler, and phase (corresp.)," *IEEE Transactions on Information Theory*, vol. 28, no. 5, pp. 817–820, Sep. 1982.

[9] K. Ho and W. Xu, "An accurate algebraic solution for moving source location using TDOA and FDOA measurements," *IEEE Transactions on Signal Processing*, vol. 52, no. 9, pp. 2453–2463, 2004.

[10] D. Mušicki, R. Kaune, and W. Koch, "Mobile emitter geolocation and tracking using TDOA and FDOA measurements," *IEEE transactions on signal processing*, vol. 58, no. 3, pp. 1863–1874, 2010.

[11] F. Guo, Y. Fan, Y. Zhou, C. Xhou, and Q. Li, *Space electronic reconnaissance: localization theories and methods*. Hoboken, NJ: John Wiley & Sons, 2014.

[12] Y. Takabayashi, T. Matsuzaki, H. Kameda, and M. Ito, "Target tracking using TDOA/FDOA measurements in the distributed sensor network," in *2008 SICE Annual Conference*, Aug 2008, pp. 3441–3446.

[13] R. Kaune, "Performance analysis of passive emitter tracking using TDOA, AOA and FDOA measurements," *INFORMATIK 2010. Service Science–Neue Perspektiven für die Informatik. Band 2*, 2010.

[14] K. C. Ho and Y. T. Chan, "Geolocation of a known altitude object from TDOA and FDOA measurements," *IEEE Transactions on Aerospace and Electronic Systems*, vol. 33, no. 3, pp. 770–783, July 1997.

[15] M. L. Fowler and X. Hu, "Signal models for TDOA/FDOA estimation," *IEEE Transactions on Aerospace and Electronic Systems*, vol. 44, no. 4, pp. 1543–1550, Oct 2008.

[16] D. Mušicki and W. Koch, "Geolocation using TDOA and FDOA measurements," in *2008 11th International Conference on Information Fusion*, June 2008, pp. 1–8.

[17] H. Yu, G. Huang, J. Gao, and B. Liu, "An efficient constrained weighted least squares algorithm for moving source location using TDOA and FDOA measurements," *IEEE transactions on wireless communications*, vol. 11, no. 1, pp. 44–47, 2012.

[18] X. Qu, L. Xie, and W. Tan, "Iterative constrained weighted least squares source localization using TDOA and FDOA measurements," *IEEE Transactions on Signal Processing*, vol. 65, no. 15, pp. 3990–4003, 2017.

[19] K. J. Cameron, "FDOA-Based Passive Source Localization: A Geometric Perspective," Ph.D. dissertation, Colorado State University, 2018.

[20] K. J. Cameron and D. J. Bates, "Geolocation with fdoa measurements via polynomial systems and ransac," in *2018 IEEE Radar Conference (RadarConf18)*. IEEE, 2018, pp. 0676–0681.

[21] M. A. Fischler and R. C. Bolles, "Random sample consensus: A paradigm for model fitting with applications to image analysis and automated cartography," *Commun. ACM*, vol. 24, no. 6, Jun. 1981.

[22] K. C. Ho, X. Lu, and L. Kovavisaruch, "Source Localization Using TDOA and FDOA Measurements in the Presence of Receiver Location Errors: Analysis and Solution," *IEEE Transactions on Signal Processing*, vol. 55, no. 2, pp. 684–696, Feb 2007.

[23] M. L. Fowler, , and S. Binghamton, "Fisher-information-based data compression for estimation using two sensors," *IEEE Transactions on Aerospace and Electronic Systems*, vol. 41, no. 3, pp. 1131–1137, July 2005.

[24] D. Adamy, *EW 103: Tactical battlefield communications electronic warfare.* Norwood, MA: Artech House, 2008.

[25] P. East, "Fifty years of instantaneous frequency measurement," *IET Radar, Sonar & Navigation*, vol. 6, pp. 112–122(10), February 2012.

[26] B. Boashash, "Estimating and interpreting the instantaneous frequency of a signal. I. Fundamentals," *Proceedings of the IEEE*, vol. 80, no. 4, pp. 520–538, April 1992.

[27] B. Boashash, "Estimating and interpreting the instantaneous frequency of a signal. II. Algorithms and applications," *Proceedings of the IEEE*, vol. 80, no. 4, pp. 540–568, April 1992.

[28] H. So, Y. Chan, Q. Ma, and P. Ching, "Comparison of various periodograms for sinusoid detection and frequency estimation," *IEEE Transactions on Aerospace and Electronic Systems*, vol. 35, no. 3, pp. 945–952, 1999.

[29] J. M. Skon, "Cramer-rao bound analysis for frequency estimation of sinusoids in noise," MIT Lincoln Laboratory, Tech. Rep. TR-727, 1986.

[30] M. A. Richards, *Fundamentals of Radar Signal Processing.* New York, NY: McGraw-Hill Education, 2014.

[31] P. M. Djuric and S. M. Kay, "Parameter estimation of chirp signals," *IEEE Transactions on Acoustics, Speech, and Signal Processing*, vol. 38, no. 12, pp. 2118–2126, Dec 1990.

[32] A. S. Kayhan, "Difference equation representation of chirp signals and instantaneous frequency/amplitude estimation," *IEEE Transactions on Signal Processing*, vol. 44, no. 12, pp. 2948–2958, Dec 1996.

[33] L. M. Huie and M. L. Fowler, "Emitter location in the presence of information injection," in *2010 44th Annual Conference on Information Sciences and Systems (CISS)*, March 2010, pp. 1–6.

Chapter 13

Hybrid TDOA/FDOA

In the previous chapters, we discussed the use of angular, temporal, and frequency estimates at multiple sensors to determine the position of an emitter, each in isolation. In this chapter, we consider the joint utilization of all three parameters. The most common arrangement discussed in the literature is the pairing of TDOA and FDOA techniques; this is generally done because both can be done with simple sensors at each receiver station and only require that either the sensors or the emitter be moving (to generate the differential Doppler needed for FDOA). The inclusion of AOA with the other two techniques requires more complex receivers at each DF station that can determine the signal's AOA, as discussed in Chapters 7 and 8. In this chapter, we will discuss the joint formulation of all three techniques, but this can easily be modified to ignore or remove one sensor modality, to analyze various paired modes (AOA/TDOA, TDOA/FDOA, and AOA/FDOA), if desired.

13.1 BACKGROUND

Within information theory, there is a concept known as *mutual information*, which quantifies the information a measurement can contain about an unknown variable (such as the sensor position) [1, 2]. In this context, additional data points are proven to never harm estimation performance, although it is possible that they add no new information, and contribute nothing to accuracy. Such is the case when two measurements are perfectly correlated or when one is fully random. With this understanding, it is an obvious choice to jointly consider as many sensor modalities as possible, hence the focus on joint estimation of position from TDOA, FDOA, and AOA. However, we must consider the possibility that additional measurements contribute nothing to the estimation performance, and represent an unnecessary increase in computational complexity.

The commonly cited argument for joint consideration of TDOA and FDOA is that the iso-measurement contours (isorange for TDOA and isodoppler for FDOA) are often perpendicular. Since geolocation errors are largest along those contours, it is often the case that errors for TDOA and FDOA are complementary. Thus, one can leverage information from TDOA to obtain an accurate measurement in one dimension and FDOA in the other. Unfortunately, the complex nature of FDOA contours does not guarantee that this is always true.

Figure 13.1 shows a simple diagram of a hybrid geolocation system, with two moving sensors that provide AOA, TDOA, and FDOA measurements. The thick solid line indicates the pair of DF measurements from the two sensors, the dotted line shows the hyperbolic arc traced by the TDOA solution, and the dash-dot line shows the isodoppler contour for the two sensors. This simple example shows that the solutions from each of these sensor modalities provides a near-perpendicular set of measurements, which underscores the utility of joint consideration. In this scenario, each measurement will provide geolocation accuracy that is best in a slightly different direction, and is likely to add significant information to the estimation process. Different sensor orientations, particularly changes in the sensor velocity directions, can have a significant impact on the direction of each contour.

In Figure 13.1, there are two sensors, each providing an angular, temporal, and frequency measurement. It is possible to construct architectures with heterogeneous sets of sensors, with some providing angular information, and some providing temporal and/or frequency information. Such a system is no different, in practice, from that in Figure 13.1, but may have significant differences in performance, given the additional diversity in sensor positions. We will continue our formulation under the assumption that all sensors provide all types of measurement, but the principles apply equally to heterogeneous systems, and the provided MATLAB® software is written to handle such cases.

13.2 FORMULATION

Geolocation from estimates of AOA, TDOA, and FDOA can be constructed via a combination of the component system equations.

$$\mathbf{z} = \begin{bmatrix} \mathbf{p}(\mathbf{x}) \\ \mathbf{r}(\mathbf{x}) \\ \dot{\mathbf{r}}(\mathbf{x}) \end{bmatrix} \tag{13.1}$$

Hybrid TDOA/FDOA

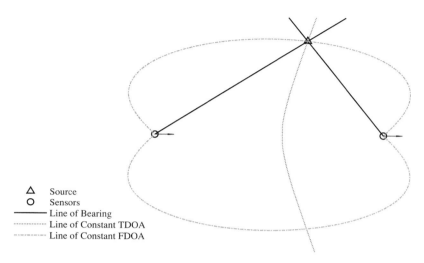

Figure 13.1 Consideration of AOA, TDOA, and FDOA jointly enables localization with only two sensors.

where $\mathbf{p(x)}$ is defined in Chapter 10, $\mathbf{r(x)}$ is defined in Chapter 11, and $\dot{\mathbf{r}}(\mathbf{x})$ is defined in Chapter 12.

If we extend the assumptions in prior chapters that each measurement is distributed as a Gaussian random vector, then probability distribution is simply

$$\zeta \sim \mathcal{N}(\mathbf{z}, \mathbf{C_z}) \qquad (13.2)$$

where $\zeta = \begin{bmatrix} \boldsymbol{\psi}^T, \boldsymbol{\rho}^T, \dot{\boldsymbol{\rho}}^T \end{bmatrix}^T$ is the noisy measurement vector and $\mathbf{C_z}$ is its covariance matrix. This covariance matrix can take many forms. If, for example, we have N sensors capable of measuring angle, time, and frequency, and the measurements are independent and identical from sensor to sensor, then the joint covariance matrix will take the form:

$$\mathbf{C_z} = \begin{bmatrix} \mathbf{C}_\psi & \mathbf{0}_{N,N-1} & \mathbf{0}_{N,N-1} \\ \mathbf{0}_{N-1,N} & \mathbf{C}_\rho & \mathbf{C}_{\rho,\dot{\rho}} \\ \mathbf{0}_{N-1,N} & \mathbf{C}_{\rho,\dot{\rho}}^T & \mathbf{C}_{\dot{\rho}} \end{bmatrix} \qquad (13.3)$$

where $\mathbf{C}_\psi$ is the $N\mathrm{x}N$ covariance matrix of all the AOA measurements, $\mathbf{C}_\rho$ is the $N-1 \times N-1$ covariance matrix of the TDOA measurements (assuming a common reference sensor), $\mathbf{C}_{\dot{\rho}}$ is the $N - 1 \times N - 1$ covariance matrix of FDOA measurements (again, assuming a common reference matrix), and $\mathbf{C}_{\rho,\dot{\rho}}$ is the $N-1 \times N - 1$ cross-covariance matrix between the TDOA and FDOA measurements. If the sensors are single-function (i.e., there are N_a AOA sensors, N_t TDOA sensors and N_f FDOA sensors, none of which are overlapping), then the covariance matrix $\mathbf{C_z}$ will be block diagonal (i.e., $\mathbf{C}_{\rho,\dot{\rho}}$ will be a matrix of zeros).

We will discuss the components and structure of $\mathbf{C_z}$ in greater detail later in this chapter.

Since the measurement matrix ζ is simply a stacked version of the component measurements, which are jointly Gaussian, the log likelihood function $\ell(\mathbf{x}|\zeta)$ is given as

$$\ell(\mathbf{x}|\zeta) = -\frac{1}{2}(\zeta - \mathbf{z}(\mathbf{x}))^H \mathbf{C_z}^{-1} (\zeta - \mathbf{z}(\mathbf{x})) \qquad (13.4)$$

The log-likelihood can be evaluated with the function `hybrid.loglikelihood` in the provided MATLAB® package, and a noise-free measurement can be quickly generated with the function `hybrid.measurement`.

Similarly, the Jacobian matrix can be defined using the components from the last three chapters.

$$\mathbf{J}(\mathbf{x}) = \begin{bmatrix} \mathbf{J_p}(\mathbf{x}) & \mathbf{J_r}(\mathbf{x}) & \mathbf{J_{\dot{r}}}(\mathbf{x}) \end{bmatrix} \qquad (13.5)$$

In our full measurement scenario (all N sensors measure TDOA, FDOA, and AOA), $\mathbf{J}(\mathbf{x})$ will have $3N - 2$ colums, and one row for each spatial dimension. In a reduced-measurement case, there will be $N_a + N_t + N_f$ columns (if there are N_a AOA sensors, N_t TDOA sensor pairs, and N_f FDOA sensor pairs). To generate the Jacobian matrix, simply utilize the function `hybrid.jacobian` in the provided MATLAB® package. This code calls `triang.jacobian`, `tdoa.jacobian`, and `fdoa.jacobian` with the appropriate inputs, and then concatenates the results.

13.3 SOLUTION

The first solution is, obviously, the ML solution, which can be estimated with a brute force solver in the provided MATLAB® function `hybrid.mlSoln`.

We plot the $\ell(\mathbf{x})$ for a series of scenarios in Figure 13.2. In each case, there are two multimode sensors providing angle, temporal, and frequency measurements, with accuracies of 10 degrees, 1 microsecond, and 100 Hz, respectively. The two sensors are marked with an o, and the true emitter position is marked with a $\triangle$. Figure 13.2(a) shows an AOA dominated case (DF accuracy improved to 5

Hybrid TDOA/FDOA

degrees). Figure 13.2(b) shows a TDOA dominated case (TDOA accuracy is 100 ns), Figure 13.2(c) shows an FDOA dominated case (FDOA accuracy is 10 Hz), and Figure 13.2(d) shows a case where all three modalities are of moderate accuracy (5 degrees, 100 ns, and 10 Hz, respectively).

These plots illustrate the relative contribution of each modality. If DF accuracy is improved from the baseline [see Figure 13.2(b)], then uncertainty at ranges beyond the target is reduced. If TDOA accuracy is improved [see Figure 13.2(c)], then cross-range errors are almost completely eliminated, but down-range uncertainty is unaffected. If FDOA accuracy is improved [see Figure 13.2(d)], then there is a significant improvement in down-range accuracy, both beyond the target and at locations closer than the target.

13.3.1 Numerically Tractable Solutions

In previous chapters, we have highlighted the utility of a both least squares and gradient descent approaches to approximate iterative solutions. The same solvers can be applied to hybrid geolocation, with the same caveats about divergence and accuracy. A convenient pair of functions are included in the provided MATLAB® package under hybrid.lsSoln and hybrid.gdSoln.

Example 13.1 Three-Channel Hybrid Solution

Consider a three-channel hybrid solution with sensors spaced 10 km abreast (side-by-side) and traveling in formation at a velocity of 100 m/s. Each platform contains a DF sensor, and time/frequency measurements for TDOA and FDOA estimation. Assuming all measurements are independent with $\sigma_\psi = 200$ mrad, $\sigma_\tau = 100$ ns, and $\sigma_f = 10$ Hz, and that TDOA and FDOA utilize a common reference sensor. Construct the measurement and estimate source location for an emitter that is 30 km in front and 30 km to the side of the central sensor, with a carrier frequency of $f_0 = 1$ GHz.

To begin, we set up the scenario geometry.

```
% Sensor and Source positions/velocities
x_sensor = 10e3 * [-1 0 1; 0 0 0];  % Spaced in x dimension
v_sensor = 100 * [0 0 0; 1 1 1];    % Moving in +y
n_sensor = size(x_sensor,2);

x_source = [1; 1]*30e3;             % Source position
x_init = [0;10e3];                  % Initial Solution
```

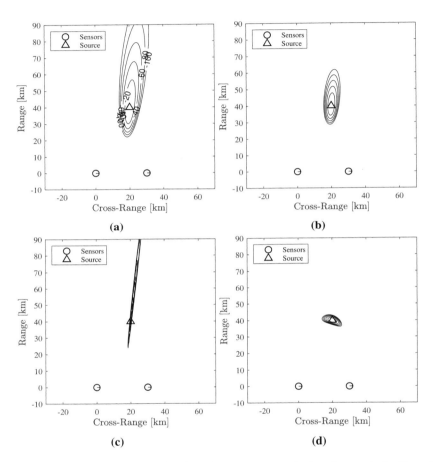

Figure 13.2 Log-likelihood at each point, given a source at the location marked with a △, and a pair of multimode sensors at the locations marked with an o: (a) DF position error is 20 degrees, TDOA error is 1 μs, and FDOA error is 100 Hz, (b) DF accuracy is improved to 1 degree, (c) TDOA accuracy is improved to 100 ns, and (d) FDOA accuracy is improved to 10 Hz.

Hybrid TDOA/FDOA

We are assuming independent measurements for the AOA, and a common reference sensor for both TDOA and FDOA, so the former is diagonal, and the latter are both a diagonal added to a common error term. We will discuss the feasibility of this (that TDOA and FDOA measurements are independent) later in this chapter.

```
% Sensor Performance
c=3e8;
f_0 = 1e9;
ang_err = .2;                   % rad
time_err = 100e-9;              % sec
rng_err = c*time_err;           % m
freq_err = 3;                   % Hz
rng_rate_err = freq_err * c/f_0; % m/s
C_psi = ang_err^2 * eye(n_sensor);
C_rdoa = rng_err^2 * (1 + eye(n_sensor-1));
C_rrdoa = rng_rate_err^2 * (1 + eye(n_sensor-1));
C_full = blkdiag(C_psi,C_rdoa,C_rrdoa);
```

Next, we generate noisy measurements based on the covariance matrix `C_full`, and apply the available solvers on each noisy data set; `hybrid.lsSoln` and `hybrid.gdSoln`. We omit the code here, because it takes the same form as examples in Chapters 10–12. An illustration of the iterative solution for both least squares and gradient descent algorithms is shown in Figure 13.3. The geolocation error, averaged across 1,000 Monte Carlo trials, is shown in Figure 13.4, as a function of the number of iterative steps computed. In Figures 13.3 and 13.4, the CRLB (derived later in this chapter) is used to plot a bound on performance. Similar to the results in Chapters 10–12, we see that the least square algorithm converges much more quickly, although both eventually approach near the CRLB. It is worth noting that these solvers require an initial position estimate, and are sensitive to errors (if it is too distant from the source, the algorithms may get trapped in a local minimum and converge to a false position estimate). The RMSE, at convergence, is just over 1 km.

Example 13.2   Heterogeneous Sensors

Consider the same scenario, but in this case the central sensor is moved to 5 km in front, and 5 km to the left ($-x$ direction), and is DF only. The remaining sensors are unmoved, and are generating time/frequency measurements for TDOA and FDOA.

The principle difference here is that the source definition and noise matrices are constructed differently

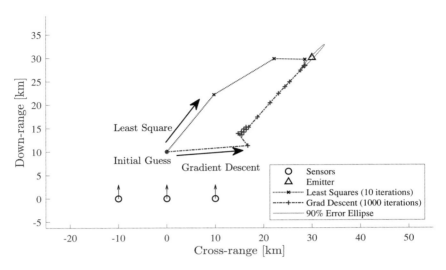

Figure 13.3 Iterative solution to Example 13.1 using both least squares and gradient descent Solvers.

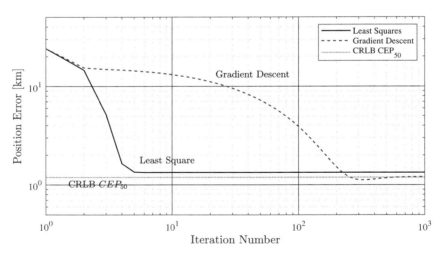

Figure 13.4 RMSE for both iterative solvers and the CRLB, for Example 13.1, as a function of the number of iterations, averaged over 1,000 Monte Carlo trials.

Hybrid TDOA/FDOA

```
% Sensor and Source positions/velocities
x_aoa = 5e3 * [-1; 1];      % DF sensor
n_aoa = 1;
x_tf = 10e3 * [-1 1; 0 0];  % TDOA/FDOA sensor pair
v_tf = 100 * [0 0; 1 1];
n_tf = 2;
```

The noise matrix is much more simply constructed since each modality (DF, TDOA, and FDOA) has a single measurement.

```
C_full = diag((ang_err*pi/180)^2,2*rng_err^2,2*rng_rate_err^2);
```

We generate the noisy measurements and geolocation error in the same manner as in prior examples. Iterative solutions for one random trial are shown in Figure 13.5, and position error as a function of the number of iterations is shown in Figure 13.6 (averaged across 1,000 Monte Carlo trials). We again note that the least square algorithm is faster to converge than gradient descent. In this case, the reduction in data available (since two sensors no longer have AOA information, and one sensor no longer has TDOA/FDOA information) leads to a modest increase in error, to slightly less than 2 km. The direction of the error ellipse is almost unchanged, despite the fact that the DF sensor has been displaced to provide an offset line of bearing. This lack of change in the error ellipse suggests that, for the example presented, DF sensors are not providing much information. If the DF sensor accuracy were to improve to, say, 1 degree or better, it is likely that this would change.

13.3.2 Other Solutions

The algorithms presented here are generic system solutions, and are intended to provide a basis for further study and analysis. They are not tuned specifically to the problems at hand, and can often be improved upon by solutions catered to a specific application. It is not common for systems to directly leverage DF sensors with TDOA and FDOA, but the latter are frequently analyzed together [3].

Ho and Xu have presented a joint TDOA/FDOA algorithm that introduces a nuisance parameter, similar to the Chan-Ho TDOA algorithm presented in Chapter 11 [4]. Similarly. Quo and Ho consider a constrained solution based on an iterative Newton's approximation, but with consideration of the rate of change of the TDOA and FDOA measurements. The additional state variables allow the simultaneous estimation of both the position and velocity of an unknown source [5]. Interested readers are encouraged to consult these sources.

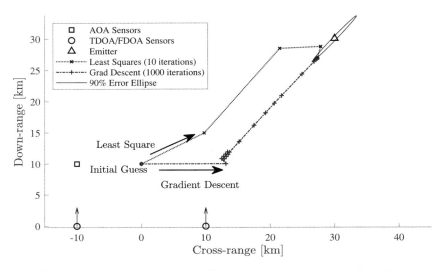

Figure 13.5 Iterative solution to Example 13.2 using both least squares and gradient descent solvers.

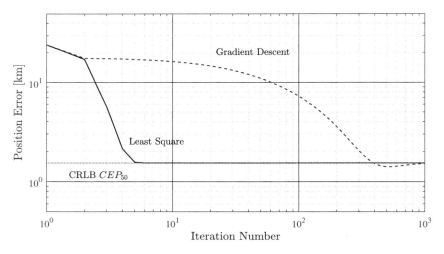

Figure 13.6 RMSE for both iterative solvers and the CRLB, for Example 13.2, as a function of the number of iterations, averaged over 1,000 Monte Carlo trials.

13.4 JOINT PARAMETER ESTIMATION

As mentioned in Chapter 12, the estimation of time of arrival or frequency of arrival of a signal is complicated by the fact that the two parameters are not independent; uncertainty about one affects the estimation accuracy of the other. In this section, we will discuss the implications of uncertainty about AOA, time of arrival, and frequency of arrival on the other parameters.

13.4.1 AOA Estimation

AOA estimation is covered in Chapters 7 and 8. It is generally considered independent of both TDOA and FDOA measurements. This is particularly true for array processing, in which the signals can first be processed for a series of steering angles, and then the data for each steering angle can be analyzed to detect signals and estimate their time and frequency content.

Thus, it is safe to assume that (a) AOA measurements are Gaussian and independent from each other, and (b) they are all independent from any TDOA and FDOA measurements. In other words, $\mathbf{C}_\psi$ is a diagonal matrix, and that the cross-terms $\mathbf{C}_{\psi,\rho}$ and $\mathbf{C}_{\psi,\dot{\rho}}$ are zero. We refer readers to Chapters 7 and 8 for suitable levels of accuracy.

13.4.2 Joint Estimation of Time/Frequency Difference

The direct estimation of frequency difference (Doppler shift for narrowband signals) has been widely studied under various conditions and various assumptions on the known/unknown nuisance parameters, such as amplitude, delay, and phase differences between the two signals, and whether one of them is noise-free (i.e., a known reference signal, or a second noisy received signal treated as a reference). Interested readers are directed to [6–12]

Estimation of time and frequency is most easily computed via the cross-ambiguity function [13, 14]. Given complex data from sensors n and m, we define the cross-ambiguity function (for some frequency shift ω and delay τ)

$$\chi_{n,m}(\omega, \tau) = \int_0^T s_n(t) s_m^*(t+\tau) e^{-j\omega t} dt \qquad (13.6)$$

The estimate of time and frequency shift is given by searching for the peak of $\chi_{n,m}(\omega, \tau)$.

$$[\widehat{\omega}_{n,m}, \widehat{\tau}_{n,m}] = \arg\max_{\omega, \tau} |\chi_{n,m}(\omega, \tau)| \qquad (13.7)$$

The classical result for Cramer-Rao bound on TDOA and FDOA between two receivers is given by Stein in 1981 [15], but was derived for a sonar signal model, in which the variance of the received signal is dependent on the unknown relative velocity. In a proper RF signal model, the delay and Doppler are deterministic (but unknown) parameters, so they have no impact on signal variance, and a much tighter bound can be formed [13, 16]. In the signal model below, delay (κ) is the digital delay in samples ($\tau = \kappa t_s$ for sample rate t_s), and doppler (ν) is in radians/sample ($\omega = \nu/t_s$).

$$\mathbf{s}_1 = \mathbf{s} + \mathbf{n}_1 \tag{13.8}$$

$$\mathbf{s}_2 = ae^{\jmath\phi}\mathbf{D}_\nu\left[\mathbf{W}^H\mathbf{D}_\kappa\mathbf{W}\right]\mathbf{s} + \mathbf{n}_2 \tag{13.9}$$

where $\mathbf{F}$ is a unitary DFT matrix, $\mathbf{D}_\kappa$ is the *delay* matrix and $\mathbf{D}_\nu$ is the *doppler* matrix, defined:

$$\mathbf{W} = \frac{1}{\sqrt{M}}\exp\left(-\jmath\frac{2\pi}{M}\mathbf{mm}^T\right) \tag{13.10}$$

$$\mathbf{D}_\kappa = \text{diag}\left\{\exp\left(-\jmath\frac{2\pi}{M}\kappa\mathbf{m}\right)\right\} \tag{13.11}$$

$$\mathbf{D}_\nu = \text{diag}\left\{\exp\left(-\jmath\nu\mathbf{m}\right)\right\} \tag{13.12}$$

$$\mathbf{m} = \left[-\frac{M}{2}, -\frac{M}{2}+1, \ldots, \frac{M}{2}-1\right]^T \tag{13.13}$$

The full unknown parameter vector is

$$\boldsymbol{\vartheta} = \left[\Re\left\{\mathbf{s}^T\right\}, \Im\left\{\mathbf{s}^T\right\}, a, \phi, \kappa, \nu\right]^T \tag{13.14}$$

It can be shown that the FIM is block diagonal, with an upper subblock for the unknown signal ($\mathbf{s}$) and the amplitude difference (a), and a lower subblock for the phase (ϕ), delay (κ), and doppler (ν). The cross-terms between these blocks are all zero, so when we invert, the lower diagonal subblock (for $\overline{\boldsymbol{\vartheta}} = [\phi, \kappa, \nu]^T$) is not impacted by the upper subblock. Thus, we can ignore the uncertainty in estimating both $\mathbf{s}$ and a. The second subblock is given [13]

$$\mathbf{F}_{\overline{\boldsymbol{\vartheta}}} = \frac{2}{a^2\sigma_1^2 + \sigma_2^2}\begin{bmatrix} E_s & -\mathbf{s}^H\widetilde{\mathbf{s}} & \widetilde{\mathbf{s}}^H\mathbf{M}\widetilde{\mathbf{s}} \\ -\mathbf{s}^H\widetilde{\mathbf{s}} & \widetilde{\mathbf{s}}^H\widetilde{\mathbf{s}} & -\Re\left\{\widetilde{\mathbf{s}}^H\mathbf{Q}_{\kappa,\nu}^H\mathbf{M}\widetilde{\mathbf{s}}\right\} \\ \widetilde{\mathbf{s}}^H\mathbf{M}\widetilde{\mathbf{s}} & -\Re\left\{\widetilde{\mathbf{s}}^H\mathbf{Q}_{\kappa,\nu}^H\mathbf{M}\widetilde{\mathbf{s}}\right\} & \widetilde{\mathbf{s}}^H\mathbf{M}^2\widetilde{\mathbf{s}} \end{bmatrix} \tag{13.15}$$

Hybrid TDOA/FDOA

where σ_1^2 and σ_2^2 are the noise variances on $s_1(t)$ and $s_2(t)$, respectively, and the modified signal vectors and auxiliary matrices are defined as

$$\bar{\mathbf{s}} = \frac{2\pi}{M} \mathbf{W}^H \mathbf{M} \mathbf{W} \mathbf{s} \tag{13.16}$$

$$\tilde{\mathbf{s}} = \mathbf{Q}_{\kappa,\nu} \mathbf{s} = \mathbf{D}_\nu \left[\mathbf{W}^H \mathbf{D}_\kappa \mathbf{W} \right] \mathbf{s} \tag{13.17}$$

$$\mathbf{M} = \text{diag}\{\mathbf{m}\} \tag{13.18}$$

$$\mathbf{Q}_{\kappa,\nu} = \mathbf{D}_\nu \left[\mathbf{W}^H \mathbf{D}_\kappa \mathbf{W} \right] \tag{13.19}$$

We note that the diagonal terms are the signal energy, a term that resembles RMS bandwidth (multiplied by signal energy), and a term that resembles RMS bandwidth (multiplied by signal energy). If we were to ignore the off-diagonal terms, then inverting the remaining matrix would result in the bounds on DF, TDOA, and FDOA accuracy from Chapters 10–12.

We wish to represent the CRLB in terms of range difference, and range-rate difference, since those are the components of $\mathbf{C}_\zeta$. We note that the conversion from $[\phi, \kappa, \nu]$ to $[r, \dot{r}]$ is given as[1]

$$\begin{bmatrix} r_{n,m} \\ \dot{r}_{n,m} \end{bmatrix} = \begin{bmatrix} 0 & ct_s & 0 \\ 0 & 0 & \frac{c}{2\pi f_0 t_s} \end{bmatrix} \begin{bmatrix} \widehat{\phi}_{n,m} \\ \kappa_{n,m} \\ \nu_{n,m} \end{bmatrix} \tag{13.20}$$

where t_s is the sampling interval, and f_0 is the signal's central frequency.

We reference Chapter 6 for the CRLB of a function of unknown variables, which leads to the CRLB

$$\mathbf{C}_{r,\dot{r}} = \begin{bmatrix} 0 & ct_s & 0 \\ 0 & 0 & \frac{c}{2\pi f_0 t_s} \end{bmatrix} \mathbf{F}_{\phi,\kappa,\nu}^{-1} \begin{bmatrix} 0 & 0 \\ ct_s & 0 \\ 0 & \frac{c}{2\pi f_0 t_s} \end{bmatrix} \tag{13.21}$$

This is the CRLB for a set of TDOA/FDOA measurements from a single set of sensors.

13.4.3 Full Covariance Matrix

The full covariance $\mathbf{C}_\mathbf{z}$ is a collection of the variance for each measurement and its cross-covariance of each pair of measurements, and is given by the block diagonal

---

[1] We convert from digital delay κ to range difference with $r_{n,m} = c\tau_{n,m} = ct_s \kappa_{n,m}$ and from digital frequency shift ν to range rate difference with $\dot{r}_{n,m} = cf_{m,n}/f_0 = (c/2\pi f_0 t_s)\nu_{m,n}$.

matrix
$$\mathbf{C_z} = \begin{bmatrix} \mathbf{C_\psi} & 0 \\ 0 & \mathbf{C_{r,\dot{r}}} \end{bmatrix} \quad (13.22)$$

The size of $\mathbf{C_\psi}$ is $N_a \times N_a$, for N_a different DF sensors. If we assume there are N_0 different TDOA/FDOA sensor pairs, then $\mathbf{C_{r,\dot{r}}}$ will have N_0 rows and columns, and will consist of four diagonal matrices.

$$\mathbf{C_{r,\dot{r}}} = \begin{bmatrix} \mathbf{C_r} & \mathbf{S_{r,\dot{r}}} \\ \mathbf{S_{r,\dot{r}}} & \mathbf{C_{\dot{r}}} \end{bmatrix} \quad (13.23)$$

where $\mathbf{C_r}$ and $\mathbf{C_{\dot{r}}}$ are the variances of the time and frequency estimates, and the diagonal matrix $\mathbf{S_{r,\dot{r}}}$ is the diagonal matrix of cross-terms between each TDOA measurement and its paired FDOA measurement.

If there is a common reference, as assumed in Chapters 11–12, then the error covariance matrix will take the same form as in prior chapters, with a block diagonal matrix and a common variance term.

$$\mathbf{C_{r,\dot{r}}} = \begin{bmatrix} \mathbf{C_r} + \sigma_{r_N}^2 & \mathbf{S_{r,\dot{r}}} + \sigma_{r_N,\dot{r}_N} \\ \mathbf{S_{r,\dot{r}}} + \sigma_{r_N,\dot{r}_N} & \mathbf{C_{\dot{r}}} | \sigma_{\dot{r}_N}^2 \end{bmatrix} \quad (13.24)$$

where the matrix terms are given by the $N-1$ unique sensors, and the scalar terms $\sigma_{r_N}^2, \sigma_{r_N,\dot{r}_N}^2$, and $\sigma_{\dot{r}_N}^2$ are the error contributions from the common reference sensor.

13.5 PERFORMANCE ANALYSIS

For geolocation performance, we again refer to the CRLB, defined as the inverse of the FIM.

$$\mathbf{C_{\hat{x}}} \geq \left[\mathbf{J}(\mathbf{x}) \mathbf{C}^{-1} \mathbf{J}^T(\mathbf{x}) \right]^{-1} \quad (13.25)$$

where the Jacobian $\mathbf{J}(\mathbf{x})$ and $\mathbf{C}$ are the full versions comprising of all AOA, TDOA, and FDOA measurements, as defined in this chapter.[2]

Figure 13.7 shows a pair of plots for an example hybrid geolocation scheme with two sensors, aligned 10 km apart in the $+x$ direction. In the first, they are moving along their baseline (in a head-to-tail configuration) at 100 m/s, and in the

---

2 Recall that the conversion from angle, range, and range-rate to position is nonlinear. Thus, even though the measurements are Gaussian, position errors are not, and they are not shown to be unbiased. The CRLB, thus, is not a strict bound (it can be violated), and cannot be used to establish efficiency of an estimator in this case. Nevertheless, it is a popular bound, and its ease of computation provides a convenient check on estimation performance.

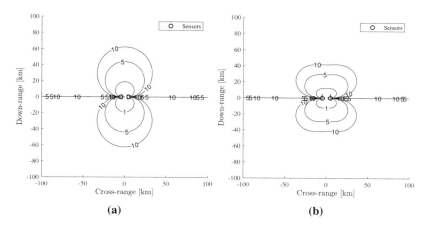

Figure 13.7 RMSE [kilometers] based on the CRLB given $\sigma_\psi = 60$ mrad, $\sigma_t = 100$ ns, $\sigma_f = 10$ Hz, $f_0 = 1$ GHz, and a sensor velocity of 100 m/s: (a) sensors moving along the baseline (head-tail) and (b) sensors moving perpendicular to baseline (side to side).

second they are moving perpendicular to their baseline (in a side-to-side configuration). In both cases, the sensors sample AOA with 3-degree accuracy, TDOA with 100-ns accuracy, and FDOA with 10-Hz accuracy. The carrier frequency is assumed to be 1 GHz (for conversion of Doppler to range rate).

The change in the shape of the estimation accuracy contours reflects the dependence of FDOA to sensor velocity. The use of TDOA and AOA tempers what would have been a loss of performance directly in front of the sensors.

Figure 13.8 plots the CRLB for a pair of test cases with just TDOA/FDOA (no AOA measurements), again with velocity of 100 m/s in the $+x$ or $+y$ direction. The lack of AOA measurements exposes the sensitivity of FDOA to sensor velocity. Figure 13.8(a) is similar to Figure 13.7(a), but does not have AOA information. The near-range performance is not greatly affected, but the accuracy beyond the 10-km contour is worse, illustrating the ability of AOA measurements to provide some accuracy when TDOA and FDOA geometries fail. Similarly, the lack of AOA measurements amplifies the dependence on sensor velocity, as seen by the significant change in performance contour between Figure 13.8(a) and Figure 13.8(b).

Figure 13.9 plots the accuracy of the same two-sensor sensor, but with TDOA and AOA measurements (no FDOA). Both TDOA and AOA are most accurate when the sensors provide a large baseline to the target; this is easily seen here by the two lobes with good performance, and the poor performance in the +/- x directions.

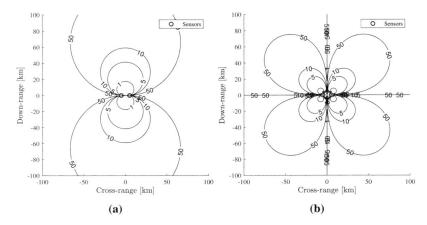

Figure 13.8 RMSE [kilometers] based on the CRLB given $\sigma_t = 100$ ns, $\sigma_f = 10$ Hz, $f_0 = 1$ GHz, and a sensor velocity of 100 m/s (No AOA sensors): (a) sensors moving along the baseline (head-tail), and (b) sensors moving perpendicular to baseline (side to side).

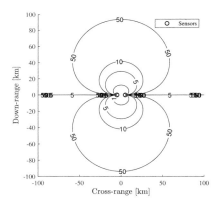

Figure 13.9 RMSE [kilometers] based on the CRLB given $\sigma_\psi = 60$ mrad, and $\sigma_t = 100$ ns (No FDOA sensors).

13.6 LIMITATIONS

We briefly discuss the limitations of TDOA and FDOA in the prior chapters, highlighting most critically the importance of accurate knowledge of the sensor positions and velocities, without which large errors are added to the time and frequency difference of arrival measurements. The inclusion of AOA data can help, primarily because the induced errors are not correlated with those induced in TDOA and FDOA measurements, and are inherently more well bounded (i.e., a 10m error in sensor position will yield an error in AOA emitter estimation on the order of 10m; for TDOA and FDOA the induced estimation errors can be much larger).

There is another set of errors that are equally problematic to all three modalities, and present a significant challenge for joint consideration, and that is multipath. Multipath echoes arrive at a different angle from the direct path, and can lead to ghost targets or to incorrect measurements of the AOA. Attempting to correctly associate TDOA, FDOA, and AOA measurements from multiple targets, particularly in the presence of multipath, is an extremely challenging problem, and is not addressed by the algorithms or performance predictions in this text.

13.7 PROBLEM SET

13.1 Plot the log-likelihood for a pair of sensors spaced 10 km apart along the x-axis, moving in formation at 100 m/s in the $+y$ direction, for a source 20-km down-range ($+y$) and 10-km cross-range ($+x$); assume $\sigma_f = 10$ Hz, $\sigma_t = 1$ μs, and $\sigma_\psi = 50$ mrad.

13.2 Repeat Problem 13.1, but adjust the headings of each sensor by either 10 degrees (a) toward each other, or (b) away from each other, and repeat the prior problem.

13.3 Repeat Example 13.1, with the central sensor 10 km behind the other two, at $x = [0, -10]$ km. Repeat 10 Monte Carlo trials to smooth the error estimates, allow the least square solution to run for 100 iterations and the gradient descent for 1,000 iterations.

13.4 Repeat Example 13.2 with the AOA sensor at $x = [0, 0]$ and the TDOA/FDOA sensor pair at $[0, 10]$ km and $[0, 20]$ km, moving in the $+y$ direction at 100 m/s. Repeat 10 Monte Carlo trials to smooth the error estimates, allow the least square solution to run for 100 iterations and the gradient descent for 1,000 iterations.

13.5 Compute the CRLB for a two-sensor solution spaced 10 km apart in the x direction, with $v = 50$ m/s in the $+x$ direction (head-to-tail). Assume $\sigma_\psi = 30$ mrad, $\sigma_t = 1\mu$ s, and $\sigma_f = 20$ Hz. Plot the CRLB for emitters as far as 100 km away with a central frequency $f = 200$ MHz.

13.6 Repeat Problem 13.5 assuming the two sensors defined are TDOA/FDOA-only (no AOA information). Add an external AOA sensor located at $[-20, 0]$ km, with $\sigma_\psi = 60$ mrad.

References

[1] C. E. Shannon, "A mathematical theory of communication," *Bell system technical journal*, vol. 27, no. 3, pp. 379–423, 1948.

[2] T. M. Cover and J. A. Thomas, *Elements of information theory*. Hoboken, NJ: John Wiley & Sons, 2012.

[3] Kimberly N. Hale, "Expanding the Use of Time/Frequency Difference of Arrival Geolocation in the Department of Defense," Ph.D. dissertation, Pardee RAND Graduate School, 2012.

[4] K. Ho and W. Xu, "An accurate algebraic solution for moving source location using TDOA and FDOA measurements," *IEEE Transactions on Signal Processing*, vol. 52, no. 9, pp. 2453–2463, 2004.

[5] F. Quo and K. Ho, "A quadratic constraint solution method for TDOA and FDOA localization," in *2011 IEEE International Conference on Acoustics, Speech and Signal Processing (ICASSP)*. IEEE, 2011, pp. 2588–2591.

[6] C. H. Knapp and G. C. Carter, "Estimation of time delay in the presence of source or receiver motion," *The Journal of the Acoustical Society of America*, vol. 61, no. 6, pp. 1545–1549, 1977.

[7] E. Weinstein and P. M. Schultheiss, "Localization of a moving source using passive array data," Naval Ocean System Center Technical Report, Tech. Rep., 1978.

[8] P. M. Schultheiss and E. Weinstein, "Estimation of differential doppler shifts," *The Journal of the Acoustical Society of America*, vol. 66, no. 5, pp. 1412–1419, 1979.

[9] M. Wax, "The joint estimation of differential delay, doppler, and phase (corresp.)," *IEEE Transactions on Information Theory*, vol. 28, no. 5, pp. 817–820, Sep. 1982.

[10] B. Friedlander, "On the cramer-rao bound for time delay and doppler estimation (corresp.)," *IEEE Transactions on Information Theory*, vol. 30, no. 3, pp. 575–580, 1984.

[11] D. Mušicki and W. Koch, "Geolocation using TDOA and FDOA measurements," in *2008 11th International Conference on Information Fusion*, June 2008, pp. 1–8.

[12] B. Friedlander, "An efficient parametric technique for doppler-delay estimation," *IEEE Transactions on Signal Processing*, vol. 60, no. 8, pp. 3953–3963, Aug 2012.

[13] A. Yeredor and E. Angel, "Joint TDOA and FDOA estimation: A conditional bound and its use for optimally weighted localization," *IEEE Transactions on Signal Processing*, vol. 59, no. 4, pp. 1612–1623, April 2011.

[14] R. Ulman and E. Geraniotis, "Wideband TDOA/FDOA processing using summation of short-time CAF's," *IEEE transactions on signal processing*, vol. 47, no. 12, pp. 3193–3200, 1999.

[15] S. Stein, "Algorithms for ambiguity function processing," *IEEE Transactions on Acoustics, Speech, and Signal Processing*, vol. 29, no. 3, pp. 588–599, June 1981.

[16] M. L. Fowler and X. Hu, "Signal models for TDOA/FDOA estimation," *IEEE Transactions on Aerospace and Electronic Systems*, vol. 44, no. 4, pp. 1543–1550, Oct 2008.

Appendix A

Probability and Statistics

In this chapter, we discuss a few select topics in probability and statistics, which arise in the detection of emitters, and estimation of their position. Interested readers are directed to [1] for basic probability theory, [2] for a handbook of distributions involving normal (or Gaussian) random variables, and [3] for detection and estimation theory. These, and other distributions relevant to RF propagation, can also be found in a publication of the International Telecommunications Union [4].

A.1 COMMON DISTRIBUTIONS

This appendix lists several common distributions that arise in detection and estimation problems. In each case, we present the *probability density function* (PDF) $f(x)$, which provides the probability that a realization of the random variable in question takes on the value x. We also present the first two central moments, known as the mean or expectation (μ) and variance (σ^2), defined as

$$\mu = E\{X\} \quad \text{(A.1)}$$
$$\sigma^2 = E\{|X - \mu|^2\} \quad \text{(A.2)}$$

where $E\{\cdot\}$ is the expectation operator, which is defined as

$$E\{g(x)\} = \int_{-\infty}^{\infty} g(x) f(x) dx \quad \text{(A.3)}$$

In the following sections, we use the notation $\sim$ to stand for "is distributed as." The PDF notation is given as $f_X(x)$ in subsequent sections, where the X term defines the distribution in use (a different symbol below for each distribution).

A.1.1 Gaussian Random Variable

The first distribution is the Gaussian, sometimes called the *normal* distribution. It is parameterized by the expectation μ and σ^2, and is written as

$$X \sim \mathcal{N}(\mu, \sigma^2) \tag{A.4}$$

$$f_\mathcal{N}(x|\mu, \sigma) = \frac{1}{\sqrt{2\pi\sigma^2}} e^{-\frac{(x-\mu)^2}{2\sigma^2}} \tag{A.5}$$

By definition, the expectation and variance of the Gaussian distribution are μ and σ^2, respectively.

$$E\{x\} = \mu \tag{A.6}$$

$$E\left\{(x-\mu)^2\right\} = \sigma^2 \tag{A.7}$$

To generate a Gaussian random variable, use the command `randn`, which generates a random sample from the *standard* normal distribution ($\mu = 0$, $\sigma^2 = 1$), and scale and offset the result.

```
x = mu + sqrt(var)*randn;
```

where `mu` is the mean μ and `var` is the variance σ^2.

A.1.2 Complex Gaussian Random Variable

A straightforward extension of Gaussian random variables is to consider complex variables. We define the complex Gaussian as the complex sum of two independent Gaussian random variables X_R and X_I, which are assumed to be independent [5, 6].[1]

$$x = x_R + jx_I \tag{A.8}$$

---

[1] There is a more complex formulation that allows dependence and requires a third parameter that captures $E\{(x-\mu)^2\}$ in addition to the variance $E\{|x-\mu|^2\}$. See [7] for formulation and details.

We parameterize the distribution with the (complex) expectation μ and (real) variance σ^2.

$$X \sim \mathcal{CN}(\mu, \sigma^2) \tag{A.9}$$

$$X_R \sim \mathcal{N}\left(\mu_R, \frac{1}{2}\sigma^2\right) \tag{A.10}$$

$$X_I \sim \mathcal{N}\left(\mu_I, \frac{1}{2}\sigma^2\right) \tag{A.11}$$

where μ_R and μ_I are the real and imaginary components of μ, respectively. Note that the real and imaginary components each have variance $\sigma^2/2$.[2] The PDF is written as

$$f_{\mathcal{CN}}(x|\mu, \sigma) = \frac{1}{\pi\sigma^2} e^{-\frac{|x-\mu|^2}{\sigma^2}} \tag{A.12}$$

By definition, the expectation and variance of the complex Gaussian distribution are μ (which can be complex) and σ^2 (which must be real-valued), respectively.

$$E\{x\} = \mu \tag{A.13}$$

$$E\{|x-\mu|^2\} = \sigma^2 \tag{A.14}$$

To generate a complex Gaussian random variable, we generate a pair of samples from the standard normal distribution, multiply one by the imaginary number (1i in MATLAB), and then scale and offset their complex sum.

```
x = mu + sqrt(var/2)*(randn+1i*randn);
```

where, once again, mu and var are the mean μ and variance σ^2, respectively. We divide the variance by two since the real and imaginary components each represent half of the total variance.

---

2   This is a notational choice and is not universal. Some sources define the components with variance σ^2 when expressing the complex Gaussian in this manner, while we define σ^2 as the sum of the component variances, which results in a slightly different form of the PDF.

A.1.3 Chi-Squared Random Variable

The next distribution is the *chi-squared* distribution, which is defined for nonnegative scalars $r \geq 0$. It is parameterized by the order k.

$$r \sim \chi_k^2 \tag{A.15}$$

$$f_{\chi^2}(r|k) = \frac{1}{2^{k/2}\Gamma(k/2)} r^{k/2-1} e^{-r/2} \tag{A.16}$$

where $\Gamma(v)$ is the gamma function.[3] The expectation and variance of a chi-squared random variable with k degrees of freedom is

$$E\{x\} = k \tag{A.17}$$

$$E\{(x-k)^2\} = 2k \tag{A.18}$$

To generate a chi-squared random variable with k degrees of freedom, use the built-in MATLAB® function chi2rnd.

Chi-squared random variables are encountered most frequently as the sum of the magnitude squared of a set of independent Gaussian random variables. If $x_l \sim \mathcal{N}(0,1)$, for $l = 0, \ldots, L-1$, then the sum of the power of each is distributed as a chi-squared random variable with L degrees of freedom.

$$z = \sum_{l=0}^{L-1} |x_l|^2 \sim \chi_L^2 \tag{A.19}$$

When the inputs are complex Gaussian, the order of the chi-squared distribution doubles from L to $2L$ (for the L real and L complex Gaussians that make up the summation), but the inputs must be distributed with variance $\sigma^2 = 2$ (so that both the real and imaginary components have unit variance).

A.1.4 Noncentral Chi-Squared Random Variable

A common generalization of the chi-squared distribution is the *noncentral* chi-squared distribution, which is parameterized by the order k and a noncentrality

---

[3] The gamma function is defined $\Gamma(z) = \int_0^\infty x^{z-1} e^{-x} dx$ [8]. When z is an integer, it reduces to the factorial $(z-1)!$.

parameter λ.

$$r \sim \chi_k^2(\lambda) \tag{A.20}$$

$$f_{\chi^\epsilon}(r|k, \lambda) = \sum_{l=0}^{\infty} \frac{e^{-\lambda/2}(\lambda/2)^l}{l!} f_{\chi^\epsilon}(r|k + 2l) \tag{A.21}$$

where $f_{\chi^\epsilon}(x|k+2l)$ is the chi-squared PDF from (A.16) with $k+2l$ degrees of freedom. The expectation and variance of the non-central chi-squared distribution are

$$E\{x\} = k + \lambda \tag{A.22}$$

$$E\left\{(x - (k + \lambda))^2\right\} = 2(k + 2\lambda) \tag{A.23}$$

To generate a non-central chi-squared random variable with k degrees of freedom, and noncentrality parameter λ, use the built-in MATLAB® function ncx2rnd.

This distribution most commonly arrives when the Gaussian variables x_l that are summed together have unit variance, but nonzero mean. In other words,

$$x_l \sim \mathcal{N}(\mu_l, 1) \tag{A.24}$$

$$z = \sum_{l=0}^{L-1} |x_l|^2 \sim \chi_N^2(\lambda) \tag{A.25}$$

where the *noncentrality parameter* λ is computed from the sum of each Gaussian random variable's squared expectation

$$\lambda = \sum_{l=0}^{N-1} |\mu_l|^2 \tag{A.26}$$

When the inputs are complex Gaussian, the form for the non-centrality parameter changes, but the order of the noncentral chi-squared distribution doubles from L to $2L$ (for the L real and L complex Gaussians that make up the summation), but the inputs must be distributed with variance $\sigma^2 = 2$ (so that both the real and imaginary components have unit variance).

A.1.5 Rayleigh Random Variable

The Rayleigh distribution arises as the amplitude of a complex Gaussian random variable with zero mean, and is parameterized by σ (the standard deviation of the Gaussian random variable that is being manipulated)

$$r \sim \mathcal{R}(\sigma) \tag{A.27}$$

$$f_\mathcal{R}(r|\sigma) = \frac{r}{\sigma^2} e^{-\frac{r^2}{2\sigma^2}} \tag{A.28}$$

The expectation and variance are given as

$$E\{r\} = \sigma\sqrt{\frac{\pi}{2}} \tag{A.29}$$

$$E\left\{\left(r - \sigma\sqrt{\frac{\pi}{2}}\right)^2\right\} = \frac{4-\pi}{2}\sigma^2 \tag{A.30}$$

To generate a Rayleigh random variable, use the built-in MATLAB® function `raylrnd`.

If $x \sim \mathcal{N}(0, \sigma^2)$ and $y \sim \mathcal{N}(0, \sigma^2)$, then the amplitude $r = \sqrt{x^2 + y^2}$ is distributed $r \sim \mathcal{R}(\sigma)$. Using our notation for complex Gaussian random variables, if $x \sim \mathcal{CN}(0, \sigma^2)$, then the amplitude $r = |x|$ is distributed $r \sim \mathcal{R}(\sigma/\sqrt{2})$.

A.1.6 Rician Random Variable

Similar to the Rayleigh random variable, the Rician random variable arises as the amplitude of a complex Gaussian random variable, but with nonzero mean. The PDF is written

$$r \sim \mathcal{R}(\nu, \sigma) \tag{A.31}$$

$$f_\mathcal{R}(r|\nu, \sigma) = \frac{x}{\sigma^2} e^{-\frac{(x^2+\nu^2)}{2\sigma^2}} I_0\left(\frac{x\nu}{\sigma^2}\right) \tag{A.32}$$

where $I_0(x)$ is the modified Bessel function of the first kind, with zero order [8], defined as

$$I_0(z) = \sum_{m=0}^{\infty} \left(\frac{1}{m!}\right)^2 \left(\frac{z}{2}\right)^{2m} \tag{A.33}$$

In MATLAB, this is computed with the function `besseli`.

The expectation and variance of a Rician distribution are given as

$$E\{r\} = \sigma\sqrt{\pi/2}L_{1/2}\left(-\nu^2/2\sigma^2\right) \quad (A.34)$$

$$E\left\{(r - E\{r\})^2\right\} = 2\sigma^2 + \nu^2 - \frac{\pi\sigma^2}{2}L_{1/2}^2\left(\frac{-\nu^2}{2\sigma^2}\right) \quad (A.35)$$

where $L_q(x)$ is the *Laguerre polynomial* [8]. An implementation of the Rician PDF and variance are provided in the MATLAB® *Statistics and Machine Learning Toolbox*, via the command

```
dist = makedist('Rician',nu,sigma);
mu = mean(dist);
variance = var(dist);
```

Implementations can also be found on the MATLAB® File Exchange. See, for example, [9].

If $x \sim \mathcal{CN}\left(\mu,\sigma^2\right)$ then $r = |x|$ is distributed as a Rician random variable with $r \sim \mathcal{R}\left(|\mu|,\sigma/\sqrt{2}\right)$.

To generate a Rician random variable with scale parameter σ and noncentrality parameter ν, the easiest choice is to use the random interface, which is supplied with the *Statistics and Machine Learning Toolbox*

```
dist = makedist('Rician',nu,sigma);
r = random(dist);
```

If that toolbox is not available, then a Rician random variable can be generated from a complex Gaussian random variable with variance $2\sigma^2$ and expectation ν.

```
xx = nu + (sqrt(2)*sigma)*(randn+1i*randn);
x = abs(xx);
```

A.2 STUDENT'S T DISTRIBUTION

The Student's T distribution is a statistical distribution first published by William Gosset (under the pen name "Student") in 1908. In the context of EW, it arises in the formulation of CFAR-matched filters; see Section 4.10 of [3] for details.

Given a set of N independent measurements, the t statistic is defined as

$$t = \frac{\overline{x} - \mu}{s/\sqrt{N}} \quad (A.36)$$

where $\bar{x}$ is the sample mean and s is the sample standard deviation, defined as

$$s^2 \triangleq \frac{1}{N-1} \sum_{i=1}^{N} (x_i - \bar{x})^2 \qquad (A.37)$$

If the measurements x_i are Gaussian distributed, then the statistic t is distributed according to the Student's T distribution with $t \sim \mathcal{T}_{N-1}$. More generally, given $X \sim \mathcal{N}(0,1)$ and $Y \sim \xi_\nu^2$, the statistic $t = X/\sqrt{Y/\nu}$ is T-distributed with ν degrees of freedom. The PDF is:

$$f_T(t|\nu) = \frac{\Gamma\left(\frac{\nu+1}{2}\right)}{\Gamma\left(\frac{\nu}{2}\right)\sqrt{\nu\pi}} \frac{1}{\left(1+\frac{t^2}{\nu}\right)^{\frac{\nu+1}{2}}} \qquad (A.38)$$

The Student's T distribution can be referenced with the functions tpdf and tcdf, for the PDF and cumulative distribution function (CDF), respectively. Both functions are provided in the *Statistics and Machine Learning Toolbox*. Generation of a random variable according to the Student's T distribution can be computed directly with the built-in MATLAB® function trnd. To manually create a variable according to Student's T, construct the components X and Y, and directly compute the test statistic:

```
x = randn(dims);
y = chi2rnd(nu,dims);
t = x./sqrt(y/nu);
```

A.3 RANDOM VECTORS

Next, consider a set of random variables x_n, $n = 0, \ldots, N-1$. If the variables are independent, then we can construct their joint PDF as simply a product of their individual PDFs

$$f_\mathbf{x}(\mathbf{x}) = \prod_{i=0}^{N-1} f_{X_i}(x_i) \qquad (A.39)$$

If, however, they are not independent, then we must apply Bayes' theorem.

$$f_\mathbf{x}(\mathbf{x}) = f_{x_0}(x_0|x_1,\ldots,x_{N-1}) f_{x_1,\ldots,x_{N-1}}(x_1,\ldots,x_{N-1}) \qquad (A.40)$$

If we carry this expansion through to all of the variables, then we have the product of marginal distributions, with each conditioned on the remaining $N - i - 1$ variables.

$$f_{\mathbf{x}}(\mathbf{x}) = \prod_{i=0}^{N-1} f_{x_i}(x_i | x_{i+1}, \ldots, x_{N-1}) \tag{A.41}$$

In many cases, as we will see below, the distribution has a much more convenient form using vectors and matrices. Just as in the case of random variables, we focus most of our attention on the first two central moments: the expectation and variance. Since $\mathbf{x}$ is a vector, its expectation is also a vector, and the variance is a matrix $\mathbf{C}$, which we refer to as the *covariance* matrix, since it specifies not just the variance of each term, but the covariance of each pair.

$$\boldsymbol{\mu} = E\{\mathbf{x}\} \tag{A.42}$$
$$\mathbf{C} = E\left\{(\mathbf{x} - \boldsymbol{\mu})(\mathbf{x} - \boldsymbol{\mu})^H\right\} \tag{A.43}$$

where $\cdot^H$ is the Hermitian operator, the complex conjugate transpose. If $\mathbf{x}$ is real, then it can be replaced with the Transpose operator $\cdot^T$.

A.3.1 Gaussian Random Vectors

A Gaussian random vector is a vector of variables x_i that each follow a Gaussian distribution, and whose pairs (x_i, x_j) are *jointly Gaussian*. The distribution of a Gaussian random vector is parameterized by the expectation $\boldsymbol{\mu}$ and covariance matrix $\mathbf{C}$

$$\mathbf{x} \sim \mathcal{N}(\boldsymbol{\mu}, \mathbf{C}) \tag{A.44}$$
$$f_{\mathcal{N}}(\mathbf{x}|\boldsymbol{\mu}, \mathbf{C}) = (2\pi)^{-N/2} |\mathbf{C}|^{-1/2} \pi^{-N/2} e^{-\frac{1}{2}(\mathbf{x}-\boldsymbol{\mu})^T \mathbf{C}^{-1}(\mathbf{x}-\boldsymbol{\mu})} \tag{A.45}$$

where $|\mathbf{A}|$ is the determinant of $\mathbf{A}$. The mean and covariance are, by definition, $\boldsymbol{\mu}$ and $\mathbf{C}$.

$$E\{\mathbf{x}\} = \boldsymbol{\mu} \tag{A.46}$$
$$E\left\{(\mathbf{x} - \boldsymbol{\mu})(\mathbf{x} - \boldsymbol{\mu})^T\right\} = \mathbf{C} \tag{A.47}$$

We can generate a random vector that follows this distribution with the randn command

```
dims = [numel(mu),1];
[V,Lam] = eig(C);        % Eigendecomposition of the Covariance matrix
Lam_sqrt = Lam.^(1/2);   % Take square root of each eigenvalue
C_sqrt = V*Lam_sqrt*V';  % Recompose square root matrix C^(1/2)
x = mu + C_sqrt*randn(dims);
```

where mu is the expectation vector μ, and C_sqrt is the square root of the covariance matrix $\mathbf{C}$.[4]

A.3.2 Complex Gaussian Random Vectors

Just as in the scalar case, we define the complex Gaussian as the complex sum of two independent Gaussian random vectors $\mathbf{x}_R$ and $\mathbf{x}_I$, which are assumed to be independent [5, 6].[5]

$$\mathbf{x} = \mathbf{x}_R + j\mathbf{x}_I \quad (A.48)$$

We parameterize the distribution with the (complex) expectation μ and (real) covariance matrix $\mathbf{C}$.

$$\mathbf{x} \sim \mathcal{CN}(\mu, \mathbf{C}) \quad (A.49)$$

$$\mathbf{x}_R \sim \mathcal{N}\left(\mu_R, \frac{1}{2}\mathbf{C}\right) \quad (A.50)$$

$$\mathbf{x}_I \sim \mathcal{N}\left(\mu_I, \frac{1}{2}\mathbf{C}\right) \quad (A.51)$$

where μ_R and μ_I are the real and imaginary components of μ, respectively.

$$f_{\mathcal{CN}}(\mathbf{x}|\mu, \mathbf{C}) = |\mathbf{C}|^{-1} \pi^{-N} e^{-\frac{1}{2}(\mathbf{x}-\mu)^H \mathbf{C}^{-1}(\mathbf{x}-\mu)} \quad (A.52)$$

The mean and covariance of $\mathbf{x}$ are, by definition, μ and $\mathbf{C}$

$$E\{\mathbf{x}\} = \mu \quad (A.53)$$

$$E\left\{(\mathbf{x}-\mu)(\mathbf{x}-\mu)^H\right\} = \mathbf{C} \quad (A.54)$$

We can generate a complex Gaussian random vector sample with the following code

---

[4] The square root $\mathbf{C}^{-1/2}$ is defined $\mathbf{C}^{-1/2}\mathbf{C}\mathbf{C}^{-1/2} = \mathbf{I}$, and is easily computed via eigendecomposition of the covariance matrix $\mathbf{C}$.

[5] There is a more complex formulation that allows dependence and requires a third parameter $\Gamma = E\{\mathbf{x}\mathbf{x}^T\}$. See [7] for formulation and details.

```
dims = [numel(mu),1];
[V,Lam] = eig(C);         % Eigendecomposition of the Covariance matrix
Lam_sqrt = (Lam/2).^(1/2); % Take square root of each eigenvalue after
                          % dividing by 2
C_sqrt = V*Lam_sqrt*V';   % Recompose square root matrix C^(1/2)
x = mu + C_sqrt*(randn(dims)+1i*randn(dims));
```

Note that the eigenvalues are first divided by two before the square root, so that the resultant matrix reflects the standard deviation of the real and imaginary components, rather than their complex sum. We then generate two random Gaussian vectors, and apply the square root of the covariance matrix to their sum.

References

[1] A. Papoulis and S. U. Pillai, *Probability, random variables, and stochastic processes*. New York, NY: McGraw-Hill Education, 2002.

[2] M. K. Simon, *Probability distributions involving Gaussian random variables: A handbook for engineers and scientists*. New York, NY: Springer Science & Business Media, 2007.

[3] L. L. Scharf, *Statistical signal processing*. Reading, MA: Addison-Wesley, 1991.

[4] ITU-R, "Report ITU-R P.1057-5, Probability distributions relevant to radiowave propagation modelling," International Telecommunications Union, Tech. Rep., 2017.

[5] B. Picinbono, "Second-order complex random vectors and normal distributions," *IEEE Transactions on Signal Processing*, vol. 44, no. 10, pp. 2637–2640, 1996.

[6] N. O'Donoughue and J. M. F. Moura, "On the product of independent complex gaussians," *IEEE Transactions on Signal Processing*, vol. 60, no. 3, pp. 1050–1063, March 2012.

[7] P. J. Schreier and L. L. Scharf, *Statistical signal processing of complex-valued data: the theory of improper and noncircular signals*. Cambridge, UK: Cambridge University Press, 2010.

[8] F. W. Olver, D. W. Lozier, R. F. Boisvert, and C. W. Clark, *NIST handbook of mathematical functions*. Cambridge, UK: Cambridge University Press, 2010.

[9] G. Ridgway. (2008) Rice/rician distribution. [Online]. Available: https://www.mathworks.com/matlabcentral/fileexchange/14237-rice-rician-distribution/

Appendix B

RF Propagation

In this appendix, we discuss the propagation of RF waves. This is an incredibly complex topic, and accurate modeling requires the inclusion of terrain data and atmospheric behavior (such as ionospheric boundary layers). To obtain extremely accurate propagation calculations, users are directed toward one of the many computational models available, such as ITM [1], TIREM [2], ITWOM [2], or the open-source SPLAT! [3]. These, other propagation models, and the underlying effects they seek to capture, are discussed in Chapter 26 of [4] and Chapter 5 of [5].

In the vast majority of situations, the level of fidelity presented by these models is either unnecessary, or not meaningful. If, for example, one is conducting a generic calculation and does not have an explicit transmitter or receiver position, then it makes little sense to generate high-resolution propagation maps that provide explicit losses at various points on the Earth's surface. In these cases, a series of simplistic models can be used, namely *free-space* attenuation, *two-ray* path loss, and *knife-edge diffraction*. Chapter 5 of [6] and Chapter 5 of [7] discuss these models at a high-level and form the basis for our approach.

For information on implementing higher fidelity models than the one described here, readers are encouraged to download one of the analytical packages referenced above, or to read the recommendations of the International Telecommunications Union, specifically [8] on the basics of propagation loss, [9] for information on ground to airborne or space-based links, [10] for discussion of diffraction effects, and [11, 12] for discussion of HF propagation (which does not generally follow the formulas derived in this appendix).

The models in this appendix are useful mostly in a system engineering context, or test and evaluation context, and not for actual receiver processing algorithms. A fielded EW receiver will often have little information about the environment in which it operates, and less about the position of emitters, so it cannot take advantage of these path loss predictions. That being said, there is increasing interest in cognitive radar, which relies on making detailed calculations about actual propagation effects based on highly accurate *digital terrain elevation data* and prior knowledge of moving targets. See, for example, [13–15]. It is possible that future developments in EW may attempt to leverage these techniques, given sufficient processing and incorporation of information from other sensors. Early efforts to incorporate cognitive techniques into EW, however, are focused on target identification and improving the detection of frequency-agile targets (or, when applied to emitters, the avoidance of jamming through adaptive frequency hopping techniques). See, for example, [16–18].

B.1 FREE-SPACE PROPAGATION

When an RF signal spreads unimpeded, the energy density along the wavefront scales with the inverse of the propagation distance squared; see Figure B.1. The path loss, which describes the ratio of the received power (by an isotropic antenna) to the emitter power (in the direction of the receiver), is given by the equation [19]

$$L_{\text{fspl}} = \left(\frac{4\pi R}{\lambda}\right)^2 \tag{B.1}$$

Free-space path loss is a valid assumption when there are no obstructions between the transmitter and receiver, and there are no dominant reflectors, such as a building or the ground, near the line of sight between the transmitter and receiver. This is mostly appropriate when either the link distance is very short, or one of either the transmitter or receiver is airborne (or space-based).

B.2 TWO-RAY PROPAGATION

For longer range situations, in the presence of a dominant reflector (such as the Earth's surface), the free-space attenuation model is no longer adequate, as the reflection of the ground begins to overlap with the direct path signal in time, causing destructive interference. See Figure B.2 for an illustration of this ground bounce path. The model was found empirically to vary with the fourth power of range, and inversely with the square of the transmitter and receiver heights. Interestingly, this

RF Propagation

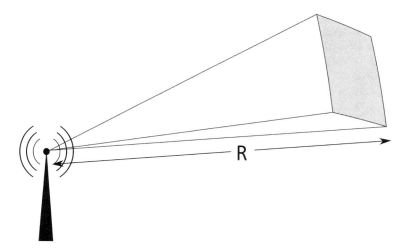

Figure B.1 Illustration of free-space path loss. The spreading is described by the surface area of a sphere with radius equal to the propagation distance.

path loss model is independent of wavelength

$$L_{2-\text{ray}} = \frac{R^4}{h_t^2 h_r^2} \tag{B.2}$$

The two-ray loss model can also be computed via a nomograph, as shown in [6].

This is one specific example of a more general class of models where loss is proportional to R^n, with some model order n that depends on the environment in question. Many models cite an order n as low as 1.5 and as high as 5 [5]. Further extensions to two-ray propagation take the Earth's curvature and surface roughness into account for either *specular* or *diffuse* reflections. See Section 5.2.4 of [5] for descriptions of the geometry involved and derivation of the resultant losses.

B.3 FRESNEL ZONE

For any terrestrial link, there is a range beyond which the free-space model is no longer appropriate and two-ray should be used. This is somewhat confusingly referred to by some as the *Fresnel zone* [6], and by others as the *turnover distance* [5].

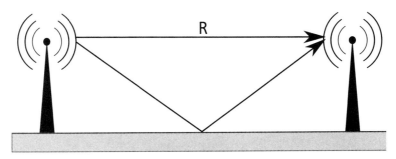

Figure B.2 In the two-ray model, reflections from the Earth's surface are the primary source of destructive interference.

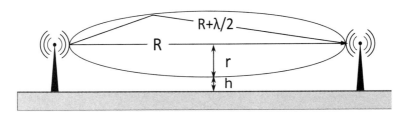

Figure B.3 Illustration of the first Fresnel zone, its radius (r), and the ground clearance (h).

We use the term *Fresnel zone distance*. Figure B.3 shows a plot of the first Fresnel zone, which is an ellipsoid centered on the line of sight between the transmitter and the receiver. The surface of this ellipsoid traces out the points where a reflection path would have a 180-degree phase shift relative to the direct path (the difference in path length between the direct path and reflection is $\lambda/2$). If there are no obstructions that interfere with this ellipsoid, and the ground is not too close, then the EM waves will propagate according to the free-space line-of-sight model. The radius of this ellipsoid at the center of the path is given as r, and the ground clearance is at that point is h. The range R at which $r/h > 60\%$ is often cited as the Fresnel zone distance, and is dependent on the height of both the transmitter and receiver off the ground, as well as the frequency.

This range can be roughly approximated as [6]

$$R_{\text{FZ}} = \frac{4\pi h_T h_R}{\lambda} \tag{B.3}$$

RF Propagation

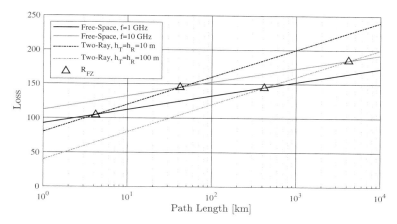

Figure B.4 Illustration of the Fresnel zone distance in relation to the free-space and two-ray path loss models. Free-space is plotted for 1 GHz and 10 GHz, two-ray for antennas 10m and 100m above the ground, and R_{FZ} is computed for the four possible pairings of the two models.

where R_{FZ} is the Fresnel zone distance (in meters), h_T and h_R are the transmitter and receiver heights (in meters), and λ is the wavelength (in meters). The form shown here is the distance at which free-space and two-ray propagation losses are equal, so it makes a convenient point for conversion between the two propagation models. Many slightly different definitions for the Fresnel zone distance can be found, relying on slightly different criteria for when it is appropriate to abandon the free-space model and apply the empirically derived two-ray model.

$$L = \begin{cases} \left(\frac{4\pi R}{\lambda}\right)^2 & R \leq R_{FZ} \\ \frac{R^4}{h_T^2 h_R^2} & R > R_{FZ} \end{cases} \quad (B.4)$$

The combined loss is plotted in Figure B.4 for several characteristic frequencies with both the free-space and two-ray loss models. The Fresnel zone distance is plotted for each pair of curves, showing the that R_{FZ} occurs where the models are equal.

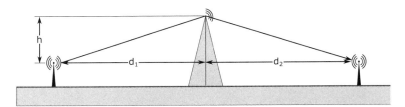

Figure B.5 Illustration of knife-edge diffraction around a peak.

An implementataion of (B.4) is provided in the MATLAB® code that accompanies this text and can be called with the function `prop.pathLoss`.[1] The value returned is in decibels, while (B.4) is defined in linear units.

B.4 KNIFE-EDGE DIFFRACTION

Knife-edge diffraction is an additional loss term that applies when there is a single obstacle (such as a mountain or tall building) between the transmitter and receiver, as shown in Figure B.5. This can also happen when a single obstacle is within the Fresnel zone ellipsoid shown in Figure B.3 but isn't blocking the line of sight path [5].

Chapter 5 of [6] discusses knife-edge diffraction, and provides a method for calculating the loss via a nomograph. Once computed, the knife-edge diffraction loss must be added to the free-space attenuation loss (not two-ray path loss, even if the link range is beyond the Fresnel zone distance). We present the equations for diffraction loss from [5], which model the same effect.

Given the geometry in Figure B.5, we compute the value ν

$$\nu = h \left(\frac{\sqrt{2}}{1 + d_1/d_2} \right) \tag{B.5}$$

where h may be the height of the line of sight path above or below the obstruction (if above, then diffraction applies when the obstruction impinges upon the first Fresnel

---

1 MATLAB® provides a number of channel models as part of the *phased array systems toolbox*, but these are geared toward modeling actual transmissions, and calculation of path loss is not explicitly returned.

zone). With this parameter, the diffraction loss is computed as

$$L_{\text{diff,dB}} = \begin{cases} 0 & \nu \leq 0 \\ 6 + 9\nu - 1.27\nu^2 & 0 < \nu < 2.4 \\ 13 + 20\log_{10}\nu & \nu \geq 2.4 \end{cases} \quad (B.6)$$

For more details on the derivation and behavior of diffraction, see Section 5.2.7 of [5] and [10].

We implement this loss (which must be added to free-space loss for a given range and frequency) in the provided utility `prop.knifeEdgeDiff`.

B.5 OTHER MODELS

Other models have been proposed to simplify the problem of path loss, including John Egli's 1957 model [20], which applies a frequency scaling factor to the two-ray model. The general R^n model uses a reference range R_0 and loss L_0 for each frequency, and scales loss for other ranges with $L_0(R/R_0)^n$. In 1988, Nicholson proposed a special case of the R^n model with n=4 and some slightly different constants [21]. Finally, a piecewise model similar to the one we employ was proposed in the 1960s, and goes by the name Longley-Rice [22]. In addition to the free-space and two-ray (line-of-sight) regions, Longley-Rice includes longer-range regions where diffraction and troposcatter are dominant.

B.6 URBAN SIGNAL PROPAGATION

The propagation of EM waves in urban environments is characterized by rich multipath and dynamic behavior. As such, it is best represented by stochastic models, rather than the deterministic path loss calculations described in the rest of this appendix. Chapter 16 of [5] discussed urban signal propagation in greater detail. These signal models are also relied on heavily in communications signal design, since the majority of cellular and wireless communications protocols must function in dense urban and indoor environments. In fact, the need for accurate modeling of communication system performance has been the driving factor in research of RF propagation in urban environments, so the majority of the models developed have been focused on frequency ranges used for cellular and indoor communications (such as GSM bands in the 800–900-MHz and 1,800-MHz regimes, IEEE 802.11 Wi-Fi in 2.4 and 5-GHz bands) [5].

The most commonly cited models for urban communications include the Okumura [23], Hata [24], and COST-231 [25] models, although many have been proposed to extend and improve them. For up-to-date recommendations, readers are directed to the relevant ITU recommendations, including [26–30]. Interested readers are referred to these citations for detailed derivations, or to Chapter 16 of [5] for a comprehensive overview.

References

[1] D. Eppink and W. Kuebler, "TIREM/SEM Handbook," Electromagnetic Compatibility Analysis Center, Annapolis, MD, Tech. Rep., 1994.

[2] S. Kasampalis, P. I. Lazaridis, Z. D. Zaharis, A. Bizopoulos, S. Zetlas, and J. Cosmas, "Comparison of longley-rice, itm and itwom propagation models for dtv and fm broadcasting," in *16th International Symposium on Wireless Personal Multimedia Communications (WPMC)*, 2013.

[3] J. Magliacane. (2014) Splat! [Online]. Available: https://www.qsl.net/kd2bd/splat.html

[4] M. Skolnik, *Radar Handbook, 3rd Edition*. New York, NY: McGraw-Hill Education, 2008.

[5] R. Poisel, *Modern Communications Jamming: Principles and Techniques, 2nd Edition*. Norwood, MA: Artech House, 2011.

[6] D. Adamy, *EW 103: Tactical battlefield communications electronic warfare*. Norwood, MA: Artech House, 2008.

[7] D. Adamy, *Practical Communications Theory, 2nd Edition*. Raleigh, NC: SciTech Publishing, 2014.

[8] ITU-R, "Rec: P.341-6, The concept of transmission loss for radio links," International Telecommunications Union, Tech. Rep., 2016.

[9] ITU-R, "Report ITU-R P.2345-0, Defining propagation model for Recommendation ITU-R P.528-3," International Telecommunications Union, Tech. Rep., 2015.

[10] ITU-R, "Rec: P.526-14, Propagation by Diffraction," International Telecommunications Union, Tech. Rep., 2018.

[11] ITU-R, "Rec: P.368-9, Ground-wave propagation curves for frequencies between 10 kHz and 30 MHz," International Telecommunications Union, Tech. Rep., 2007.

[12] ITU-R, "Rec: P.1148-1, Standardized procedure for comparing predicted and observed HF skywave signal intensities and the presentation of such comparisons," International Telecommunications Union, Tech. Rep., 1997.

[13] J. R. Guerci, R. M. Guerci, M. Ranagaswamy, J. S. Bergin, and M. C. Wicks, "CoFAR: Cognitive fully adaptive radar," in *2014 IEEE Radar Conference*, May 2014, pp. 0984–0989.

[14] G. E. Smith, "Cognitive radar experiments at the ohio state university," in *2017 Cognitive Communications for Aerospace Applications Workshop (CCAA)*, June 2017, pp. 1–5.

[15] J. R. Guerci, "Cognitive radar: A knowledge-aided fully adaptive approach," in *2010 IEEE Radar Conference*, May 2010, pp. 1365–1370.

[16] S. Kuzdeba, A. Radlbeck, and M. Anderson, "Performance metrics for cognitive electronic warfare - electronic support measures," in *MILCOM 2018 - 2018 IEEE Military Communications Conference (MILCOM)*, Oct 2018, pp. 1–9.

[17] S. You, M. Diao, and L. Gao, "Deep reinforcement learning for target searching in cognitive electronic warfare," *IEEE Access*, vol. 7, pp. 37 432–37 447, 2019.

[18] Ryno Strauss Verster and Amit Kumar Mishra, "Selective spectrum sensing: A new scheme for efficient spectrum sensing for ew and cognitive radio applications," in *2014 IEEE International Conference on Electronics, Computing and Communication Technologies (CONECCT)*, Jan 2014, pp. 1–6.

[19] ITU-R, "Rec: P.525-3, Calculation of free-space attenuation," International Telecommunications Union, Tech. Rep., 2016.

[20] J. J. Egli, "Radio propagation above 40 mc over irregular terrain," *Proceedings of the IRE*, vol. 45, no. 10, pp. 1383–1391, 1957.

[21] D. L. Nicholson, *Spread spectrum signal design: LPE and AJ systems*. Rockville, MD: Computer Science Press, Inc., 1988.

[22] A. G. Longley and P. L. Rice, "Prediction of tropospheric radio transmission loss over irregular terrain, a computer method," ESSA Technical Report ERL 79–ITS 67, NTIS Access No. 676-874, Tech. Rep., 1968.

[23] Y. Okumura, E. Ohmori, T. Kawano, and K. Fukuda, "Field strength and its variability in vhf and uhf land-mobile radio service," *Record of the Electronic Communication Laboratory*, vol. 16, pp. 825–873, 1968.

[24] M. Hata, "Empirical formula for propagation loss in land mobile radio services," *IEEE transactions on Vehicular Technology*, vol. 29, no. 3, pp. 317–325, 1980.

[25] K. Löw, "A comparison of cw-measurements performed in darmstadt with the cost-231-walfisch-ikegami model," Rep. COST 231 TD, Tech. Rep., 1991.

[26] ITU-R, "Report ITU-R P.1406-2, Propagation effects relating to terrestrial land mobile and broadcasting services in the VHF and UHF bands," International Telecommunications Union, Tech. Rep., 2015.

[27] ITU-R, "Report ITU-R P.530-17, Propagation data and prediction methods required for the design of terrestrial line-of-sight systems," International Telecommunications Union, Tech. Rep., 2017.

[28] ITU-R, "Report ITU-R P.1238-9, Propagation data dn prediction methods for the planning of indoor radiocommunication systems and radio local area networks in the frequency range 300 MHz to 100 GHz," International Telecommunications Union, Tech. Rep., 2017.

[29] ITU-R, "Report ITU-R P.1410-5, Propagation data and prediction methods required for the design of terrestrial broadband radio access systems operating in a frequency range from 3 to 60 GHz," International Telecommunications Union, Tech. Rep., 2012.

[30] ITU-R, "Report ITU-R P.1546-5, Method for point-to-area predictions for terrestrial services in the frequency range 30 MHz to 3000 MHz," International Telecommunications Union, Tech. Rep., 2013.

Appendix C

Atmospheric Absorption

This appendix covers the losses experienced by propagating EM waves that are a result of the absorption of energy by the materials that they pass through. These losses are typically broken into three parts:

- Losses due to absorption by gases (including water vapor), L_g;
- Losses due to absorption by rain, L_r;
- Losses due to absorption by clouds and fog, L_c.

In each of these cases, the loss can be expressed by a coefficient γ that is expressed in terms of decibels per kilometer and is multiplied by a distance across which each loss term is applied D (in kilometers) to obtain the total loss L (in decibels).

$$L_{atm} = L_g + L_r + L_c \qquad (C.1)$$
$$= \gamma_g D_g + \gamma_r D_r + \gamma_c D_c \qquad (C.2)$$

Figure C.1 illustrates the different ranges involved, with a path traversing D_g km of atmosphere, D_r of which is impacted by rain and D_c of which is impacted by clouds and fog.

For another high-level overview, see Section 5.8 of [1]. These losses are derived in greater detail in Chapter 5 of [2] and Chapter 19 of [3]. The details below are loosely taken from these sources, but rely more heavily on the recommendations of the International Telecommunications Union [4–10].

Similar to Appendix B, the loss terms computed here are most directly applicable to system engineering and to performance prediction. In a fielded system,

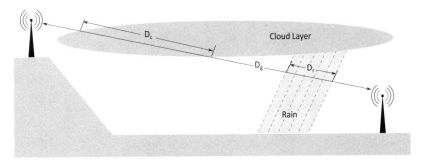

Figure C.1 Illustration of path lengths associated with absorption by gases (D_g), rain (D_r), and clouds (D_c).

we are unlikely to have detailed knowledge of the atmospherics necessary to accurately compute these losses, and their impact may be minimal compared with propagation losses in Appendix B or the noise terms in Appendix D. That being said, an understanding of the expected atmospheric losses in certain scenarios can inform the likely and guaranteed detection ranges.

C.1 LOSSES DUE TO ABSORPTION BY GASES

Absorption of EM energy by gases and water vapor is represented by an absorption coefficient γ_g given in units of decibels per kilometer. The total loss, L_g is computed as

$$L_g = \gamma_g D_g \tag{C.3}$$

where D_g is the distance of propagation (one-way) in kilometers. The coefficient γ_g varies with frequency, atmospheric pressure, temperature, and water vapor density. It is dominated at RF frequencies by the absorption from atmospheric oxygen (γ_o) and water vapor (γ_w). A suitable approximation, presented for frequencies below 350 GHz, is provided in [4], which we implement in the accompanying MATLAB code, in the function `atm.gaslossCoeff`.

We plot the atmospheric loss coefficient as a function of frequency for the standard atmosphere (defined later in this appendix) at several altitudes (0, 10, and 20 km above sea level) in Figure C.2.[1]

---

[1] In general, losses are more severe at lower altitudes.

Atmospheric Absorption

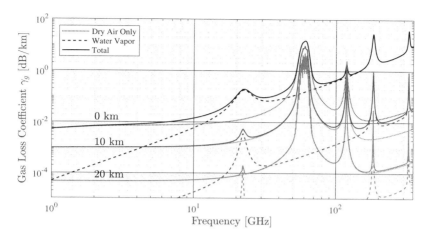

Figure C.2 Gas loss coefficient at $h = 0$, 10, and 20 km altitude for a standard atmosphere, as a function of frequency.

For frequencies below 1 GHz, the loss is below .005 dB/km (often much lower); representing a cumulative loss of 0.5 dB every 100 km. This is small enough to be negligible for all but the longest transmissions. At 10 GHz, γ_g increases to 0.015 dB/km, and at 20 GHz to 0.1 dB/km. At these frequencies, the loss over 100 km is 1.5 dB and 10 dB, respectively. For frequencies above 20 GHz, the losses are even more significant.

C.2 LOSSES DUE TO ABSORPTION BY RAIN

Absorption due to falling rain is described by [5] with a piecewise model that takes in the elevation angle and polarization angle of the transmission. The resultant loss is governed by two parameters k and α, and the path-length through the rain

$$L_r = \gamma_r D_r \tag{C.4}$$
$$\gamma_r \triangleq k R^\alpha \tag{C.5}$$

where R is the rain rate (expressed in millimeters per hour). If multiple periods of rain are encountered, then the loss for each section should be computed, and then the losses added (in decibel space).

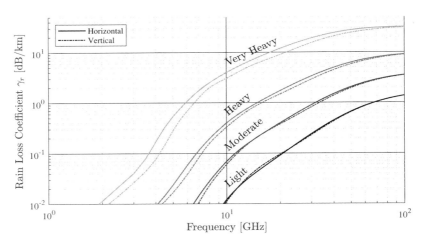

Figure C.3 Rain loss coefficient for four representative rain levels defined in Table C.1.

Table C.1
Representative Rainfall Conditions Used in Figure C.3

Rain Condition	Light rain	Moderate rain	Heavy rain	Very heavy rain
Rainfall Rate	1 mm/hr	4 mm/hr	16 mm/hr	100 mm/hr

The equations are implemented in the provided MATLAB package in the function atm.rainlossCoeff.

We plot the rain loss coefficient in Figure C.3 as a function of frequency for both horizontal and vertically polarized signals with a path elevation angle of $0°$ for several characteristic rain levels, defined in Table C.1.

For light rain, losses are negligible below frequencies of 10 GHz. As the rainfall rate increases, the minimum affected frequency decreases to as low as 2 GHz for very heavy rainfall (100 mm/hr). Looking at the different rainfall rates for 10 GHz, the loss varies from as low as .01 dB/km in light rain to as high as 4 dB/km, illustrating the significant impact that rainfall can have on absorption loss.

Table C.2
Representative Cloud Densities Used in Figure C.4

Fog Thickness	Density
600-m visibility	.032 g/m^3
120-m visibility	.32 g/m^3
30-m visibility	2.3 g/m^3

C.3 LOSSES DUE TO ABSORPTION BY CLOUDS AND FOG

Absorption by clouds and fog is dictated primarily by the density of the fog, represented in grams of liquid water per cubic meter.[2] The loss coefficient is given

$$\gamma_c = K_l M \tag{C.6}$$

where M is the fog density (grams per cubic meter), and K_l is a parameter defined by the model in [4]. We implement computation of that parameter in the provided MATLAB® code, in the function `atm.foglossCoeff`.

The resultant loss coefficient is plotted in Figure C.4 for a series of cloud densities, parameterized by the visibility of fog with a given cloud density in Table C.2.

The impact of dense clouds and fog can be seen as low as 1 GHz, while light fog has negligible impact at frequencies below approximately 20 GHz. In all cases, the effect is roughly linear with frequency (in log-log space), with impact at 10 GHz between .02 dB/km for moderate visibility and slightly more than .1 dB/km for low visibility (30 m).

C.4 STANDARD ATMOSPHERE

The standard reference atmosphere, as defined by [7] is computed with the MATLAB® script `atm.standardAtmosphere`, and defines a piecewise temperature, pressure, and water vapor density, as a function of height. For simplicity, we do not reproduce them here.

---

[2] This is based on the assumption that all water droplets are small compared with the wavelength λ, and is valid for frequencies below 200 GHz [6].

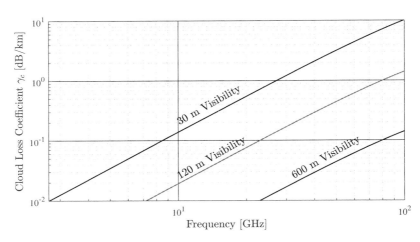

Figure C.4 Gas loss coefficient for three representative cloud densities defined in Table C.2.

C.5 WRAPPER FUNCTION

For simplicity, we also apply a wrapper function that can be called with a set of atmospheric and path variables, which returns a single loss L_{atm}. It is called with the script

```
atmStruct = struct('T',T,'p',p,'e',e,'R',r,'M',m);
L = atm.calcAtmLoss(f,Dg,Dr,Dc,atmStruct,pol_angle,el_angle);
```

where Dg, Dr, and Dc are the path lengths for each of the three components of loss, atmStruct is a MATLAB® struct containing the necessary atmospheric parameters, including the rainfall rate and cloud density (both of which are assumed equal to zero if they are not supplied). The final arguments pol_angle and el_angle are the polarization and elevation angle of the signal, which are defined and called for in the rain absorption model.

C.6 ZENITH ATTENUATION

Zenith attenuation, the total atmospheric loss between the Earth's surface and the edge of the atmosphere for a transmission at zenith (directly up), we can compute

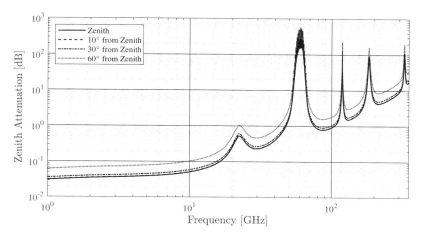

Figure C.5 Zenith loss attenuation due to gas loss

the loss at various altitudes, and then accumulate. This behavior is provided in the function `atm.calcZenithLoss`.

This same approach can be used to calculate the total loss along slanted paths by first dividing the path into segments, and then accumulating the total loss in each segment. A simple approximation is provided within the function `atm.calcZenithLoss`. If an optional third argument, the zenith angle, is provided then the loss is computed along a straight line through the atmosphere, which is pointed `zenith_angle` radians from nadir (straight up).

This result is plotted for frequencies between 1 and 350 GHz in Figure C.5 for zenith loss (straight up), as well as slant range paths at $10°$, $30°$, and $45°$ from nadir. The estimation used here is limited in that it assumes a straight-line path from the starting position to the edge of the atmosphere. In reality, the path will curve as the density of the atmosphere varies, so these results should be taken as an estimate, particularly for large zenith angles.

C.7 MATLAB® TOOLBOXES AND MODEL FIDELITY

The provided functions are loosely related to built-in MATLAB® functions `gaspl`, `rainpl`, and `fogpl`, which are part of the *Phased Array Systems Toolbox* (both derive from the relevant ITU recommendations). It is important to note that the loss due to atmospheric gases is highly dependent on altitude. Both our utility and the built-in MATLAB® functions operate at a single altitude. To consider slanted paths, consider breaking it into subpaths at various altitudes, and adding the component losses to determine the total loss of the slanted path.

References

[1] D. Adamy, *EW 103: Tactical battlefield communications electronic warfare*. Norwood, MA: Artech House, 2008.

[2] L. V. Blake, *Radar Range-Performance Analysis*. Norwood, MA: Artech House, 1986.

[3] M. Skolnik, *Radar Handbook, 3rd Edition*. New York, NY: McGraw-Hill Education, 2008.

[4] ITU-R, "Rec: P.676-11, Attenuation by atmospheric gases in the frequency range 1–350 GHz," International Telecommunications Union, Tech. Rep., 2016.

[5] ITU-R, "Rec: P.838-3, Specific attenuation model for rain for use in prediction methods," International Telecommunications Union, Tech. Rep., 2005.

[6] ITU-R, "Rec: P.840-7, Attenuation due to clouds and fog," International Telecommunications Union, Tech. Rep., 2017.

[7] ITU-R, "Rec: P.835-6, Reference Standard Atmospheres," International Telecommunications Union, Tech. Rep., 2017.

[8] ITU-R, "Rec: P.836-6, Water vapour: surface density and total columnar content," International Telecommunications Union, Tech. Rep., 2017.

[9] ITU-R, "Rec: P.837-7, Characteristics of precipitation for propagation modelling," International Telecommunications Union, Tech. Rep., 2017.

[10] ITU-R, "Rec: P.839-4, Rain height model for prediction methods," International Telecommunications Union, Tech. Rep., 2013.

Appendix D

System Noise

Noise is a reality for any and all receivers. Even in the absence of interfering man-made signals, there is galactic background radiation and RF emissions from celestial objects, the atmosphere, and the ground (sometimes called *sky noise*) as well as emissions from excitation of the circuitry and components within and near the RF system in question, referred to as *thermal noise*. A detailed discussion of the environmental sources of noise, including a discussion of the thermodynamics and quantum mechanics that give rise to thermal noise, can be found in Chapter 4 of [1]. Other sources, including man-made impulsive noise sources, are detailed in Sections 2.8 and 2.9 of [2]. A concise set of recommendations is also found in [3].

In this appendix, we briefly review the additive white noise model, which is by far the most commonly cited approach. We then discuss briefly the generalization of this model to handle frequency-dependent (nonwhite) noise; introduce the dominant components of sky noise; and impulsive noise sources.

D.1 ADDITIVE WHITE GAUSSIAN NOISE

The dominant source of noise in a receiver is *thermal noise*, which "is caused by the thermal agitation of electrons in am imperfect conductor" [1]. This fact leads to an immediate observation, which is that cooling a system can result in lower noise levels, and corresponding improvements in sensitivity. Put simply, the noise level is the product of the temperature (in degrees Kelvin) and Boltzmann's constant ($k = 1.38 \times 10^{-23}$ Watt-seconds/degree Kelvin).

$$N_0 = kT \tag{D.1}$$

The standard temperature, sometimes called the *reference temperature* is 290K, which corresponds to a *room temperature* of 17°C. In (D.1), N_0 is the noise power spectral density (watt-seconds, or equivalently, watts per hertz). To obtain the total noise power received, we also multiply by the bandwidth of the receiver, referred to as the *noise bandwidth* B_n

$$N = kTB_n \tag{D.2}$$

where N is the total noise power in watts. This represents an ideal system. In reality, RF components are lossy and non-ideal. These losses lead not only to a decrease in received signal strength, but also to an increase in thermal noise, since the energy absorbed by the transmission line is eventually re-radiated. To account for this, an empirical *noise figure* is measured for each component. The *noise figure* is defined as the excess noise measured above that predicted by (D.2).

$$F_n = \frac{\widetilde{N}}{kTB_n} \tag{D.3}$$

where $\widetilde{N}$ is the measured noise power (in watts), and F_n is expressed in linear units. Sometimes it is more convenient to express the noise figure in decibel units, NF, defined as

$$\text{NF} = 10\log_{10}(F_n) \tag{D.4}$$

With this, the noise power can be computed in either linear or decibel units

$$N = kTBF_n \tag{D.5}$$
$$N\,[\text{dB}] = 10\log_{10}(kTB) + \text{NF} \tag{D.6}$$

This value (in decibels) can be computed with the provided MATLAB function `noise.thermal_noise`.

Section 5-2 of [4] contains a detailed discussion of noise figure measurements, and the calculation of effective noise figure for modern receivers (which comprise many components). Some representative noise figure values described include a value between 10 and 20 dB for traveling wave tube (TWT) amplifiers, 6 dB for more modern solid-state power amplifiers, less than 1.5 dB for lossy antennas (as low as 0.05 dB for very low loss antennas), and a noise figure equivalent to the loss factor for passive network components, such as splitters. The combined noise figure for a receiver system will include contributions from each stage (e.g., antennas, amplifiers, filters, phase shifters, and analog-to-digital converters).

D.2 COLORED NOISE

If, for some reason, the noise spectrum is colored (not uniform across frequency), we replace the product kTB with an integral

$$N = (F_n - 1) \int_{f_0-B/2}^{f_0+B/2} N_0(f) df \tag{D.7}$$

where $N_0(f)$ is the frequency-dependent power spectral density, and the passband of the receiver is $f_0 + [-B/2, B/2]$.

D.3 SKY NOISE

While the thermal noise explained above is often considered sufficient for general modeling purposes, additional sources of noise should be considered if high fidelity is desired. Blake [1] and the ITU [3] present models for several of the noise sources. We summarize a few of the dominant sources here. These additional noise sources can be included as modifications of the system temperature T from (D.1)

$$T_{eff} = T + T_C + T_A + T_G \tag{D.8}$$

where each of these terms refers to the contribution of a different source of noise. T is the thermal noise temperature (290K), T_C is the noise temperature from cosmic sources (including the Sun and Moon), T_A is from noise emitted by Earth's atmosphere, and T_G is from noise emitted by the ground.[1]

To compute the effective noise level, we call thermal_noise with an optional third argument, to specify the external noise temp.

```
N = noise.thermal_noise(bw, nf, Tc);
```

It is not always necessary to include sky noise; in fact in many cases it has minimal impact on the resultant noise figure (an increase of less than 1 dB). Figure D.1 plots this relationship, with the increase in noise level ($kB(T_{eff} - T)$) as a function of the increase in noise temperature ($T_{eff} - T$). From Figure D.1, we can see that if the combined contribution of all external noise sources is less than 35 K, then the increase in the noise level will be less than 0.5 dB. If it is less than

---

[1] The noise temperatures in (D.8) include implicit modifiers, such as the fraction of integrated antenna gain in the direction of each noise source. These terms are explicitly discussed in [1].

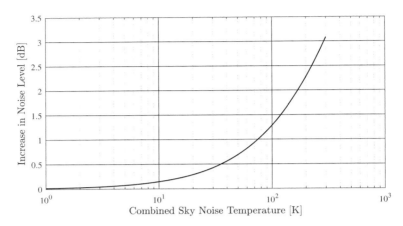

Figure D.1 The noise level increase is a function of the external noise temperature, with less than a 0.5-dB increase in total noise power if the combined external noise temperature is less than 35K.

10 K, then the increase is less than 0.1 dB. To effect a 3-dB increase in the noise level (a doubling of the noise power), the external noise sources need to have a combined noise temperature of 290K.

D.3.1 Cosmic Noise

The first term, T_C, is the collection of all sources of EM radiation outside of the Earth's atmosphere, with the exception of the Sun. The galactic background level contributes 2.7K to noise temperature, while the contribution of discrete sources (nearby stars) follows a frequency-dependent (and variable) noise level. The integrated equation is

$$T_C = \frac{\alpha_c}{L_A} \left[T_{100} \left(\frac{100}{f_{MHz}} \right)^{2.5} + 2.7 \right] \quad (D.9)$$

where f_{MHz} is the carrier frequency (in megahertz), T_{100} is the galactic noise temperature at the reference frequency (100 MHz), α_c is the fraction of the total integrated antenna pattern pointed above the horizon, and L_A is the average atmospheric loss across different portions of the sky. Blake argues that $\alpha = 0.95$ is a good approximation for most ground-based radars. T_{100} fluctuates between 500 and

18,650K, with a geometric mean value of 3,050K. Estimation of L_A should follow the approaches in Appendix C. We use loss L_A at an elevation angle of $45°$ as an approximate to the average loss across the sky.

Above 2 GHz, cosmic noise is essentially negligible, unless the antenna is pointed directly at a bright object, such as the Moon or the Sun [3]. Cosmic noise levels can be estimated with the provided MATLAB® function noise.cosmic_noise_temp.

D.3.1.1 Solar Noise

Because the Sun is a centralized noise source in the sky, we can model its effects separately from cosmic noise. Just as in the case of cosmic noise, the effective noise temperature includes antenna pattern and atmospheric loss terms.

$$T_s = \frac{\alpha_s}{L_A} \widetilde{T}_s \tag{D.10}$$

Absent solar flares, the Sun's temperature $\widetilde{T}_s$ varies from 10^5K at 1 GHz, to as high as 10^6K below 300 MHz, and to below 10^4K above 10 GHz. Since the Sun is a discrete object in the sky, this contribution can often be ignored if the antenna is not pointed at the Sun, since α_s will represent an antenna sidelobe. Given this fact, Blake presents a more convenient form where α_s (the fraction of the antenna's receive pattern that is pointed at the Sun) is replaced with G_s (the average antenna gain in the direction of the Sun)

$$T_s = 4.75 \times 10^{-6} \frac{G_s \widetilde{T}_s}{L_A} \tag{D.11}$$

If the sidelobes G_s are below 0 dBi, then even without atmospheric loss, the effective noise temperature contributed by the Sun will be less than 4.75K, roughly on the order of the galactic background radiation contribution. To include the effects of the Sun, we provide an optional argument, G_sun to the function cosmic_noise_temp.

```
Tc = noise.cosmic_noise_temp(f,rx_alt,alpha_c,G_sun);
```

where G_sun is G_s, the average antenna gain pointed toward the Sun (in decibels relative to an isotropic antenna—dBi).

D.3.1.2 Lunar Noise

We include lunar noise in the same manner as solar noise. Noise from the Moon has a constant level of approximately 200K between 1 GHz and 100 GHz [3], and

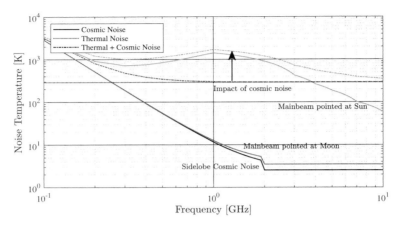

Figure D.2 Cosmic noise temperature and thermal noise temperature as a function of frequencies for baseline (ignoring solar and lunar noise), and cases where a 30-dBi antenna mainlobe is pointed at either the Sun or the Moon.

follows (D.11), with substitution of the antenna gain toward the Moon (G_m) instead of G_s, and $\widetilde{T}_m = 200K$ instead of $\widetilde{T}_s$. In the provided code, this can be called by providing an optional argument to the function cosmic_noise_temp

```
Tc = noise.cosmic_noise_temp(f,rx_alt,alpha_c,G_sun,G_moon);
```

where G_moon is G_m (in dBi).

D.3.1.3 Combined Cosmic Noise

Figure D.2 plots the cosmic noise temperature for frequencies between 100 MHz and 10 GHz for both a generic scenario, and a 30-dBi mainlobe pointed at either the Moon or the Sun. A 30-dBi mainlobe pointed at the Moon has an almost negligible impact, resulting in a total noise temperature from cosmic sources that is less than 35K for frequencies above 600 MHz. The impact of pointing a 30-dBi mainlobe at the Sun is much more intense, with noise levels above 35K at frequencies above 10 GHz. From this, we conclude that cosmic (including solar and lunar) noise is minimal (< 30K) above 600 MHz and can be safely ignored above 1 GHz (< 10K), unless the antenna is pointed at the sun.

D.3.2 Atmospheric Noise

The atmosphere is a contributor of noise in the bands where it is absorptive. This is because the atmosphere's components absorb, and then reradiate, energy. This behavior can be modeled as the noise from a lossy transmission line [1]. The noise temperature T_A can thus be computed with the atmospheric temperature T_t and the absorptive loss L_A along the antenna's primary pointing angle

$$T_a = \alpha_a T_t \left(1 - \frac{1}{L_A}\right) \tag{D.12}$$

where T_t and L_A are averaged over the propagation path from the antenna to the edge of the atmosphere (in the antenna mainlobe). α_a is the fraction of the antenna's receiver energy over the path defined. Given that atmospheric noise is coming from all angles, we can assume $\alpha_a = 1$ without significant loss of precision. See Blake [1] for more exact forms of (D.12) that include integrating along the propagation path. The temperature can be taken from the standard atmosphere, described in Appendix C, as can the atmospheric loss.

We plot the atmospheric noise temperature T_a as a function of elevation angle and frequency in Figure D.3. This was computed by calculating the atmospheric loss to the edge of the Earth's atmosphere (L_A) for each elevation angle, and computing the average temperature between ground level and the edge of the atmosphere (100-km altitude) for the standard atmosphere defined in Appendix C, resulting in the approximation $T_t = 228$K. Under these assumptions, atmospheric noise is minimal (< 30K) in almost all scenarios below 20 GHz and can be ignore (< 10K) below 10 GHz. Above 20 GHz, the noise level is significantly effected by elevation angle and the absorption bands (where there is significant atmospheric loss).

D.3.3 Ground Noise

The Earth itself, like all celestial objects, is an emitter of radiation. The Earth's thermal temperature is reasonably approximated with $T_t = 290$ K. The noise contribution of the Earth can be computed as [1]

$$T_G = \frac{\Omega_G G_G \mathcal{E} T_t}{4\pi} \tag{D.13}$$

where Ω_G is the percentage of the antenna's field of view that is occupied by the Earth (which can be approximated with π steradians [1]), G_G is the average antenna

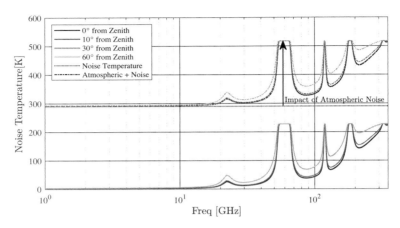

Figure D.3 Atmospheric noise temperature as a function of frequency and elevation angle for a ground-based receiver with $\alpha_a = 1$ and $T_t = 228$ K.

gain in the direction of the Earth (which Blake argues can be approximated with a value between 0.1 and 0.5), and $\mathcal{E}$ is the surface emissivity of the Earth, which is (by definition) between 0 and 1. For perfectly reflective surfaces, $\mathcal{E} = 0$, while for perfectly opaque surfaces, $\mathcal{E} = 1$. Given these values, and typical approximations, the contribution of ground noise is as low as 7.3K (when $G_G = 0.1$) and as high as 36K (when $G_G = 0.5$).

Figure D.4 plots the ground noise temperature (T_g) as a function of the average antenna gain pointed toward the ground (G_G), under the assumption that $\Omega_G = \pi$ steradians, $\mathcal{E} = 1$, and $T_t = 290$K. In this case, ground noise (in isolation) is minimal (< 30K) when $G_G < -4$ dBi, and can be safely ignored (< 10K) when $G_G < -8$ dBi.

D.4 URBAN (MAN-MADE) NOISE

Signal reception in urban environments must contend with a tremendous amount of man-made interference, including from consumer electronics (such as microwave ovens, vehicle ignition systems, and cellular phones) and industrial equipment (such as welders and utility-scale power generation and distribution). See Section 2.8 from [2] for a discussion of these sources and Section 2.9 of [2] for a mathematical formulation of their impact to signal detection.

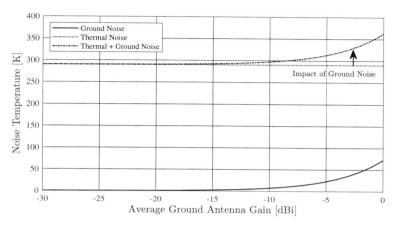

Figure D.4 Ground noise temperature as a function of average antenna gain pointed toward the ground, when $\Omega_G = \pi$ steradians, $\mathcal{E} = 1$, and $T_t = 290$K.

References

[1] L. V. Blake, *Radar Range-Performance Analysis*. Norwood, MA: Artech House, 1986.

[2] R. Poisel, *Modern Communications Jamming: Principles and Techniques, 2nd Edition*. Norwood, MA: Artech House, 2011.

[3] ITU-R, "Rec: P.372-13, Radio Noise," International Telecommunications Union, Tech. Rep., 2016.

[4] Avionics Department, *Electronic Warfare and Radar Systems: Engineering Handbook*, 4th ed. China Lake, CA: Naval Air Warfare Center Weapons Division, 2013.

About the Author

Dr. Nicholas A. O'Donoughue is a senior engineer with the RAND Corporation, where he provides radar, communications, and electronic warfare expertise to a broad array of defense studies through RAND's Federally Funded Research and Development Centers. He earned his B.Cp.E. degree from Villanova University in 2006, and both the M.S. and Ph.D. in electrical engineering from Carnegie Mellon University in 2009 and 2011, respectively. Following the completion of his Ph.D., Dr. O'Donoughue worked in the Airborne Radar Systems and Techniques Group at MIT Lincoln Laboratory from 2012–2015, where he focused on the analysis and development of electronic warfare techniques for airborne ground surveillance radar systems.

Dr. O'Donoughue was an instructor for several iterations of the "Build a Radar Course" offered through MIT's Professional Education Program from 2013–2015, and offered a course on radar signal processing at Tufts University in the Spring of 2015.

He is a recipient of the National Defense Science and Engineering Graduate Fellowship, the Dean Robert D. Lynch Award from the Villanova University Engineering Alumni Society, and the Computer Engineering Outstanding Student Medallion from Villanova University. He has published more than 40 technical journal and conference papers, including two that were chosen as best student paper. Nicholas is a senior member of the IEEE, and he is a member of both the Tau Beta Pi and Eta Kappa Nu engineering honor societies.

Index

absorption, 309–316
    cloud loss, 309, 313
    gas loss, 309–311
    rain loss, 309, 311–312
    reference atmosphere, 313
    zenith, 314–315
analog-to-digital converter, 73, 74
angle of arrival, *see* direction finding
atmospheric loss, *see* absorption
attenuation, *see* propagation

bandwidth, 2–4, 7, 8, 14, 29–32, 49–50, 53, 55, 71, 75
beamforming, 145–146
beamwidth, 154
beat frequencies, 74
Bhattacharyya bound, *see* performance bound, Bhattacharyya bound

Cramér-Rao lower bound, *see* performance bound, Cramér-Rao lower bound (CRLB)

detection, 1–4, 13–26

constant false alarm rate (CFAR), 24–25
cross-correlation, 58–62
cyclostationary, 55–58
energy detector, 33, 55
false alarm, 17, 23
hypothesis test, 13
    binary, 17–21, 33
    composite, 21–24
likelihood ratio, 17
log likelihood ratio, 19
matched filter, 22
missed detection, 17
square law detector, 35
sufficient statistic, 15–16, 21, 23
threshold, 17–20, 34–35
direction finding, 1, 4, 101–103, 107–141, 145–178
    Adcock, 112–114
    amplitude-based, 108–116
    beamscan, 163–164
    Capon beamformer, 164
    Doppler, 120–126

330   INDEX

  estimation of signal parameters via rotational invariance technique (ESPRIT), 171
  minimum variance distortionless response (MVDR), 164–165
  minimum-norm, 171
  monopulse, 137–141
  MUSIC, 169–171
  phase interferometry, 127–132
  reflector, 114–116
  robust Capon beamformer, 167–168
  subspace, 168–171
  Watson-Watt, 117–120
distribution, 287–297
  Chi-squared, 290
  Chi-squared, non-central, 290–291
  Complex Gaussian, 288–289, 296–297
  Gaussian, 17, 288, 295–296
  likelihood function, 14
  multivariate, 294–297
  Rayleigh, 292
  Rician, 292–293
  Student's T, 25, 293–294

effective radiated power (ERP), 41
electronic intelligence (ELINT), 7
emissions control (EMCON), 9
error
  bias, 186
  circular error probable (CEP), 190–192
  confidence interval, 186–188
  ellipse, 186–190
  root mean square (RMSE), 95, 101, 102, 134–136, 177, 178, 209, 273
  root mean square error (RMSE), 193
  spherical error probable (SEP), 192–193
estimation
  Bayes, 89–91
  convex, 92
  least square, 91–92
  maximum likelihood, 85–87, 159–163
  minimum variance unbiased (MVUB) estimator, 88
  nuisance parameters, 97–98, 129, 131, 225, 255, 275, 277
  tracking, 92–93

FDOA, 5
Fisher Information Matrix, 96, 119, 193–194
Fisher information matrix, 96–100, 111, 112, 120, 124, 130, 174, 211, 233, 278, 280
Frequency difference of arrival (FDOA)
  gradient descent estimate, 253
  isodoppler, 248
  iterative least square estimate, 251–253
  maximum likelihood estimate, 250–251
  range-rate difference, 248–249
frequency hopping, 8, 49
  bandwidth, 71–72, 75–77
  period, 75–77

INDEX

gain
    array, 153–154
geolocation, 1, 5–6
grating lobes, 127, 129, 152, 153

hybrid geolocation
    gradient descent estimate, 271–275
    iterative least square estimate, 271–275
    maximum likelihood estimate, 270–271

irregular arrays, 146–147

Jacobian, 194

knife-edge diffraction, *see* propagation, knife-edge

line of bearing, *see* direction finding
Link-16, 8
local oscillator, 73

matrix inversion lemma, 97, 112, 130
minimum detectable signal (MDS), 43, 44
modulation
    binary phase shift keying (BPSK), 50
    phase shift keying (PSK), 50, 54

noise, 13–14, 317–324
    atmospheric, 323
    bandwidth, 318
    cosmic, 320–322
    figure, 14, 318
    galactic background, 13
    ground, 323–324
    power spectral density (PSD), 318
    reference temperature, 318
    sky, 317, 319–324
    thermal, 13, 317–318
Normal distribution, *see* distribution, Gaussian

path loss, *see* propagation
performance bound
    Bhattacharyya bound, 100
    Cramér-Rao lower bound, 95
    Cramér-Rao lower bound (CRLB), 5, 100, 126, 131, 134–137, 172–178, 193–195, 211, 233–234, 257–258
    Ziv-Zakai Bound (ZZB), 5
    Ziv-Zakai bound (ZZB), 100
power spectral density (PSD), 30, 32
probability
    covariance matrix, 295
    density function (PDF), 287
    expectation, 287, 295
    variance, 287
propagation, 37–41, 299–306
    atmospheric loss, 40
    computational models, 305
    link equation, 37
    path loss
        free space, 38, 299–300
        Fresnel zone, 39, 301–303
        knife edge diffraction (KED), 299, 304–305
        two-ray, 38, 299–301
pulse descriptor word, 229, 239

radar warning receivers, 1
receiver

complex, 73
digital, 72–73
dwell period, 76–77, 79, 80
frequency hopping, 75
hopping period, 78
intermediate frequency, 73–75
noise figure, 41, 44
scan period, 75–77
superheterodyne, 74
RMS bandwidth, 230–231, 256
RMS duration, 256

Signal-to-noise ratio (SNR), 2, 23, 34, 42, 100
spread spectrum, 8
direct sequence spread spectrum (DSSS), 50
frequency hopping spread spectrum (FHSS), 50
processing gain, 53
spreading gain, *see* spread spectrum, processing gain
steering vector, 148–149

taper, 154–157

TDOA, 5
Time difference of arrival (TDOA)
cross-correlation, 228, 230–232
gradient descent estimate, 224–225
isochrone, 219
iterative least square estimate, 222–224
leading edge detection, 228, 230
maximum likelihood estimate, 222
peak detection, 228–230
range difference, 220–221
triangulation, 5
centroid, 203
gradient descent estimate, 207–208
incenter, 203
iterative least square estimate, 205–207
maximum likelihood estimate, 203–205

Ziv-Zakai bound, *see* performance bound, Ziv-Zakai bound

The Artech House Electronic Warfare Library

Dr. Joseph R. Guerci, Series Editor

Activity-Based Intelligence: Principles and Applications, Patrick Biltgen and Stephen Ryan

Advances in Statistical Multisource-Multitarget Information Fusion, Ronald P. S. Mahler

Antenna Systems and Electronic Warfare Applications, Richard A. Poisel

Electronic Intelligence: The Analysis of Radar Signals, Second Edition, Richard G. Wiley

Electronic Warfare for the Digitized Battlefield, Michael R. Frater and Michael Ryan

Electronic Warfare in the Information Age, D. Curtis Schleher

Electronic Warfare Receivers and Receiving Systems, Richard A. Poisel

Electronic Warfare Signal Processing, James Genova

Electronic Warfare Target Location Methods, Richard A. Poisel

Emitter Detection and Geolocation, for Electronic Warfare, Nicholas A. O'Donoughue

EW 101: A First Course in Electronic Warfare, David L. Adamy

EW 103: Tactical Battlefield Communications Electronic Warfare, David L. Adamy

EW 104: EW Against a New Generation of Threats, David L. Adamy

Foundations of Communications Electronic Warfare, Richard A. Poisel

High-Level Data Fusion, Subrata Das

Human-Centered Information Fusion, David L. Hall and John M. Jordan

Information Warfare and Organizational Decision-Making, Alexander Kott, editor

Information Warfare and Electronic Warfare Systems, Richard A. Poisel

Information Warfare Principles and Operations, Edward Waltz

Introduction to Communication Electronic Warfare Systems, Richard A. Poisel

Introduction to Electronic Defense Systems, 3rd Edition, Filippo Neri

Introduction to Modern EW Systems, 2nd Edition, Andrea De Martino

Knowledge Management in the Intelligence Enterprise, Edward Waltz

Mathematical Techniques in Multisensor Data Fusion, Second Edition, David L. Hall and Sonya A. H. McMullen

Military Communications in the Future Battlefield, Marko Suojanen

Modern Communications Jamming Principles and Techniques, Richard A. Poisel

Practical ESM Analysis, Sue Robertson

Principles of Data Fusion Automation, Richard T. Antony

RF Electronics for Electronic Warfare, Richard A. Poisel

Stratagem: Deception and Surprise in War, Barton Whaley

Statistical Multisource-Multitarget Information Fusion, Ronald P. S. Mahler

Tactical Communications for the Digitized Battlefield, Michael Ryan and Michael R. Frater

Target Acquisition in Communication Electronic Warfare Systems, Richard A. Poisel

For further information on these and other Artech House titles, including previously considered out-of-print books now available through our In-Print-Forever® (IPF®) program, contact:

Artech House	Artech House
685 Canton Street	16 Sussex Street
Norwood, MA 02062	London SW1V 4RW UK
Phone: 781-769-9750	Phone: +44 (0)20-7596-8750
Fax: 781-769-6334	Fax: +44 (0)20-7630-0166
e-mail: artech@artechhouse.com	e-mail: artech-uk@artechhouse.com

Find us on the World Wide Web at: www.artechhouse.com